Ocean Outpost
The Future of Humans Living Underwater

Erik Seedhouse

Ocean Outpost

The Future of Humans Living Underwater

 Springer

Published in association with
Praxis Publishing
Chichester, UK

Dr Erik Seedhouse, F.B.I.S., As.M.A.
Milton
Ontario
Canada

SPRINGER–PRAXIS BOOKS IN POPULAR SCIENCE
SUBJECT *ADVISORY EDITOR*: Stephen Webb, B.Sc., Ph.D.

ISBN 978-1-4419-6356-7   e-ISBN 978-1-4419-6357-4
DOI 10.1007/978-1-4419-6357-4

Springer New York Dordrecht Heidelberg London

Library of Congress Control Number: 2010927367

© Springer Science+Business Media, LLC 2011

Cover design: Jim Wilkie
Project copy editor: Christine Cressy
Typesetting: BookEns, Royston, Herts., UK

Printed on acid-free paper

Springer is a part of Springer Science+Business Media (www.springer.com)

Contents

viii   **Contents**

Preface

In an era in which satellite photographs chart even the most remote land masses in astonishing detail, the vast majority of our planet lies unrevealed beneath the ocean. In this watery wilderness, an environment every bit as inaccessible as space, humans have rarely ventured more than a few hundred meters below the waves. At a time at which most people think of space as the final frontier, we should remind ourselves that a great deal of unfinished business remains here on Earth and as robots crawl on the surface of Mars, we should remember that most of our own planet has never been seen with human eyes.

One of the greatest scientific and technological achievements of the twenty-first century will be to cast a light on the eternal darkness of the deep ocean. Life, say the experts, began in the ocean, and if the way many people spend their vacations is any indication, there are few activities we enjoy more than revisiting our submerged origins. Whether cavorting with dolphins, harassing sharks from the protection of titanium cages, or photographing exotic aquatic species through the portholes of tourist submarines, humans have a natural affinity with what lies beneath the surface of the ocean. But despite having had the technology to establish permanent settlements under the ocean for more than five decades, of the 100 billion humans who have ever lived, not a single one has lived permanently underwater.

Ocean Outpost traces the future of man underwater, describing how technology will allow humans to adapt to a permanent life underwater. This book also unveils the challenges that will be faced by twenty-first-century aquatic pioneers and, ultimately, colonists, in what may in due course prove to be one of the greatest human adventures in history.

To realize the goal of a permanent human presence underwater, a wealth of new technologies will need to be developed and qualified, including new manned and unmanned submersibles, advanced propulsion systems, underwater rescue systems, decompression methods, and revolutionary physiological intervention strategies ranging from liquid ventilation to artificial gills. Some of the technology development and medical intervention will require quantum leaps in innovation, while others may be nothing short of radical, appearing to be more science fiction than science. Step by step, *Ocean Outpost* describes how the technology will evolve, how crews will be selected and trained, and what a typical underwater mission will entail. The book also chronicles the frontiers of underwater technology that will eventually take humans into the midst of a world we could once only guess at.

This book is dedicated to those who accept the dangers and challenges of furthering the human dream to establish a permanent human presence under the ocean and to all those who support them.

Acknowledgments

In writing this book, the author has been fortunate to have had five reviewers who made such positive comments concerning the content of this publication and to Clive Horwood and his team at Praxis for guiding this book through the publication process. The author also gratefully acknowledges all those who gave permission to use many of the images in this book, especially Jeff Parker of Ambient Pressure Diving, world-record No Limits freedivers Herbert Nitsch, Patrick Musimu, and Tanya Streeter, scientists Dr. Robert Freitas, Dr. Phil Nuytten, Alan Bodner, and Erika Shagatay, Will Kohnen, President/CEO of SEAmagine Hydrospace Corporation, open-circuit scuba world record-holder, Mark Ellyat, and Hawkes Ocean Technologies.

The author also expresses his deep appreciation to Christine Cressy, whose attention to detail and patience greatly facilitated the publication of this book, to Jim Wilkie for creating the cover of this book, and to Stewart Harrison, who sourced several of the references that appear in *Ocean Outpost*.

Once again, no acknowledgment would be complete without special mention of our cats, Jasper and MiniMach, who provided endless welcome distraction and entertainment.

About the author

Erik Seedhouse is an aerospace scientist whose ambition has always been to work as an astronaut. After completing his first degree in Sports Science at Northumbria University, the author joined the legendary 2nd Battalion the Parachute Regiment, the world's most elite airborne regiment. During his time in the "Para's", Erik spent 6 months in Belize, where he was trained in the art of jungle warfare and conducted several border patrols along the Belize–Guatemala border. Later, he spent several months learning the intricacies of desert warfare on the Akamas Range in Cyprus. He made more than 30 jumps from a Hercules C130 aircraft, performed more than 200 abseils from a helicopter, and fired more light anti-tank weapons than he cares to remember!

Upon returning to the comparatively mundane world of academia, the author embarked upon a Master's degree in Medical Science at Sheffield University. He supported his Master's degree studies by winning prize money in 100-km ultradistance running races. Shortly after placing third in the World 100 km Championships in 1992 and setting the North American 100-km record, the author turned to ultradistance triathlon, winning the World Endurance Triathlon Championships in 1995 and 1996. For good measure, he also won the inaugural World Double Ironman Championships in 1995 and the infamous Decatriathlon, the world's longest triathlon, an event requiring competitors to swim 38 km, cycle 1,800 km, and run 422 km. Non-stop!

Returning to academia once again in 1996, Erik pursued his PhD at the German Space Agency's Institute for Space Medicine. While conducting his PhD studies, he still found time to win Ultraman Hawaii and the European Ultraman Championships as well as completing the Race Across America bike race. Due to his success as the world's leading ultradistance triathlete, Erik was featured in dozens of magazines and television interviews. In 1997, *GQ* magazine nominated him as the "Fittest Man in the World".

In 1999, Erik decided it was time to get a real job. He retired from being a professional triathlete and started his post-doctoral studies at Vancouver's Simon Fraser University's School of Kinesiology. While at Simon Fraser University, Erik established the Simon Fraser Freediving Program in association with Performance

Freedivers. In 2005, the author worked as an astronaut training consultant for Bigelow Aerospace in Las Vegas and wrote *Tourists in Space*, a training manual for spaceflight participants. He is a Fellow of the British Interplanetary Society and a member of the Aerospace Medical Association. Recently, he was one of the final 30 candidates of the Canadian Space Agency's Astronaut Recruitment Campaign. Erik currently works as an aerospace project manager – a job that includes such duties as acceleration training officer for the Canadian Forces, manned centrifuge operator, and flight director for manned hyperbaric operations. In his spare time, he also works as a manned spaceflight consultant, triathlon coach, and author. He plans to travel into space with one of the private spaceflight companies.

In addition to being a triathlete, skydiver, pilot, and author, Erik is an avid scuba-diver and has logged more than 200 dives in more than 20 countries. His favorite dive destinations are Moorea and Tasmania. His favorite diving movies include *The Big Blue*, by Luc Besson, the director's cut of *The Abyss*, by James Cameron, and the classic 1989 Norwegian film, *Dykket* (*The Dive*), by Tristan DeVere Cole. His favorite underwater science fiction novels include *Starfish* by Peter Watts and *OceanSpace* by Allen Steele. *Ocean Outpost* is his sixth book. When not writing, he spends as much time as possible in Kona on the Big Island of Hawaii and at his real home in Sandefjord, Norway. Erik lives with his wife and two cats on the Niagara Escarpment in Canada.

Figures

Tables

Panels

Abbreviations and acronyms

AAT	Aquatic Ape Theory
ABE	Autonomous Benthic Explorer
ABS	American Bureau of Shipping
ADME	Absorption, Digestion, Metabolism, and Elimination
ADS	Atmospheric Diving Suit
AIP	Air Independent Propulsion
AMS	Atmospheric Monitoring System
ARDS	Acute Respiratory Distress Syndrome
ARS	Air Revitalization System
ATB	Acoustic Transponder Beacon
ATCS	Active Thermal Control System
AUV	Autonomous Underwater Vehicle
BCD	Buoyancy Control Device
CCR	Closed Circuit Rebreather
CMBB	Center for Marine Biotechnology and Biomedicine
CNS	Central Nervous System
COTS	Commercial off the Shelf
CSA	Canadian Space Agency
CT	Computed Tomography
CWS	Caution and Warning System
DCS	Decompression Sickness
DDIA	Dubai Development and Investment Authority
DND	Department of National Defence
DSRV	Deep Submergence Rescue Vehicle
DWA	Drop Weight Assembly
EU	European Union
FDA	Federal Drug Administration
FPB	Fixed Positive Buoyancy
HA	Hydroxyapatite
HOT	Hawkes Ocean Technologies
HPNS	High Pressure Nervous Syndrome

HUD	Head-Up Display
IND	Investigational New Drug
ISS	International Space Station
JAMSTEC	Japan Marine Science and Technology Center
LARS	Launch and Recovery System
LiOH	Lithium Hydroxide
LBL	Long-baseline
LSS	Life Support System
MBT	Main Ballast Tank
MDR	Mammalian Diving Reflex
MMS	Minerals Management Service
MOD	Maximum Operating Depth
MRDF	Marine Resources Development Foundation
NAUI	National Association of Underwater Instructors
NDT	Non-Destructive Testing
NMRI	Naval Medical Research Institute
NOAA	National Oceanic Atmospheric Administration
NRC	National Research Council
OBS	Optical Backscatter
OWTT	One Way Travel Time
PADI	Professional Association of Diving Instructors
PEM	Polymer Electrode Membrane
PFC	Perfluorocarbon
PI	Principal Investigator
PIT	Pilot in Training
OCS	Outer Continental Shelf
OWC	OceanWorks International Corporation
RHOV	Replacement Human-Occupied Vehicle
RLV	Residual Lung Volume
ROV	Remotely Operated Vehicle
SAS	Sub Aviator Systems
SIO	Scripps Institute of Oceanography
SMS	Seafloor Massive Sulphide
SOR	Strategic Oil Reserve
SRC	Submarine Rescue Chamber
SRDRS	Submarine Rescue Diving and Recompression System
TCCS	Trace Contaminant Control Subassembly
THCS	Thermal and Humidity Control System
TLC	Total Lung Capacity
UNCW	University of North Carolina Wilmington
VBS	Variable Ballast System
VBT	Variable Ballast Tank
WHOI	Woods Hole Oceanographic Institution
WMS	Waste Management System

Section I

Diving

In June 1913, the *Regina Margherita*, flagship of the Italian Navy, anchored in Picadia Bay, Karpathos, in the Aegean Sea. In a heavy storm, the ship dragged its anchor, which eventually broke and was lost in 77 m of ocean. Since it would have been a disgrace for the captain to return to the home port without the anchor, several attempts were made to salvage it. After many unsuccessful days spent trying to recover the anchor, one of the divers died, the Italian Naval Archives describing his death as the result of a blackout.

In desperation, the captain sought help from a group of Greek sponge fishermen renowned for their diving abilities and offered a substantial reward to the diver who could recover the ship's anchor. Several divers offered to help, among them a rather feeble-looking fisherman by the name of Georghios (Yorgos) Haggi Statti, who boasted he could hold his breath for 7 min and would have no trouble diving down and recovering the anchor. At 1.75 m tall and weighing 60 kg, Yorgos did not look the part of a diver capable of descending to 77 m on one breath of air and because of his sickly appearance, the captain was initially quite skeptical and requested that the ship's doctor perform a medical examination.

The results of the examination did little to convince the captain, since it was discovered that Yorgos was suffering from pulmonary emphysema, a debilitating condition of the respiratory system. Secondary symptoms associated with this condition do not help a breath-hold diver, since both the resting heart rate and the breathing rate are elevated. In the case of Yorgos, his breathing rate was measured at between 20 and 22 breaths per minute (a normal rate being about 12 breaths per minute), and his heart rate at between 80 and 90 beats per minute (compared with a normal rate of 70 beats per minute). However, Yorgos did appear to have one advantage, since his lack of an auditory membrane permitted him to dive without having to equalize using his hand. The doctors recommended that given the illness, Yorgos should not be allowed to dive. Fortunately, Yorgos, not being a fan of current medical practice, chose to ignore the doctors, and began training for the dive anyway. Within a period of 4 days, he made 21 practice dives to depths of between 45 and 84 m – performances that astounded the doctors. These depths are particularly remarkable given that in 1960, the Cousteau team declared that in their estimation, a depth of 55 m was the absolute limit for human breath-hold diving.

Yorgos dove down to 77 m on three occasions and succeeded in passing a rope through the anchor's eye that allowed the crew to retrieve it. It should be noted that during these dives, Yorgos wore neither a mask nor fins. For his trouble, he received five pounds of sterling and given permission to fish with dynamite – a practice usually reserved for the Italian Navy. He also passed into the annals of breath-hold-diving legend and became the subject of numerous articles, his story being mentioned in books written by future breath-hold-diving athletes.

Strangely, each account of Yorgos's feat sidesteps the obvious question: How did he do it? The answers can be found in testimonies and statements made by the doctors who examined him, which are preserved in the archives of the Italian Navy in Rome. The explanation lies in a simple but, at the time, ingenious technique used by Yorgos. He descended in the vertical head-up position after first tying a rock weighing about 45 kg to his ankles. This ballast allowed him to descend to the bottom at a phenomenal rate. Once there, he simply cut the rope with a knife, passed it through the anchor's eye, and allowed the surface crew to pull him back to the surface.

Today, athletes still practice a version of the diving technique pioneered by Yorgos in a discipline known as No Limits, which is the subject of the first chapter of this book. Section I begins by describing the challenges of modern No Limits freedivers as they set their sights on dives to 300 m on just one breath of air. Chapter 2 ventures deeper by exploring the world of technical and saturation divers and identifies the factors that determine life or death for divers working at extreme depths. Section I concludes by assessing future science and technology that will be required by divers seeking to extend the human diving envelope by venturing into previously impenetrable depths.

1

No Limits Freediving

"The challenges to the respiratory function of the breath-hold diver[1] are formidable. One has to marvel at the ability of the human body to cope with stresses that far exceed what normal terrestrial life requires."

Claes Lundgren, Director, Center for Research and
Education in Special Environments

A woman in a deeply relaxed state floats in the water next to a diving buoy. She is clad in a figure-hugging wetsuit, a dive computer strapped to her right wrist, and another to her calf. She wears strange form-hugging silicone goggles that distort her eyes, giving her a strange bug-eyed appearance. A couple of meters away, five support divers tread water near a diving platform, watching her perform an elaborate breathing ritual while she hangs onto a metal tube fitted with two crossbars. A few meters below the buoy, we see that the metal tube is in fact a weighted sled attached to a cable descending into the dark-blue water.

Her eyes are still closed as she begins performing a series of final inhalations, breathing faster and faster. Photographers on the media boats snap pictures as she performs her final few deep and long hyperventilations, eliminating carbon dioxide from her body. Then, a thumbs-up to her surface crew, a pinch of the nose clip, one final lungful of air, and the woman closes her eyes, wraps her knees around the bottom bar of the sled, releases a brake device, and disappears gracefully beneath the waves. The harsh sounds of the wind and waves suddenly cease and are replaced by the effervescent bubbling of air being released from the regulators of scuba-divers.

Bright beams of light illuminate her as she descends on the sled at an alarming rate, hanging on like a ribbon in the wind, her 1-m-long carbon fiber fins fluttering like the wings of an insect. The platform soon disappears from sight as she plunges faster, faster, into an immiscible abyss, the crystalline light fading at a rate of more than 2 m per second. A marker at 30 m is reached in a matter of seconds as two safety divers watch from a distance. Four atmospheres. At this depth, the pressure

---

[1] The more common term is "freediver", which is used throughout this chapter.

would squeeze a balloon to only one-fourth the size it had been on the surface. All is silent except for the occasional click-click of a nearby dolphin witnessing this surreal aquatic performance.

Her depth gauge reads 60 m. Seven atmospheres. Everything around her is a dark blue. She spots two more safety divers, one of their spotlights trained on the aquatic human descending the plumb line to the darkness below. The divers look clear and sharp but their colors appear washed out into bluish grays like an old black-and-white film shown late at night on a bad television set. She hits a thermocline, the ocean's deep icebox layer, and the water temperature plunges to 7°C, immediately numbing her lips and cheeks. Dark blue shades to black. Points of light brilliance occasionally dance in front of her eyes. Phosgene. This is deep. Very deep. Like a human torpedo, the aquatic human closes on the next marker at an astonishing rate. 150 m. This is below the crush depth of many World War II submarines. 170 m. Already, she can feel the strain in her respiratory system as carbon dioxide starts to build in her muscles and organs and is taken away in her bloodstream to her lungs.

She appears serene as she clutches the sled to the infinity below, apparently oblivious of the remarkable changes occurring in her body. Her lungs have already been crushed to the size of an orange and her heart is barely beating. The blood shift occurring at these depths means her spleen has pumped out a mass of extra blood cells, her blood vessels have collapsed, and blood has been forced out of her limbs into the space where her lungs should be. The elapsed time since leaving the surface is now 1 min and 10 sec.

180 m. She is experiencing a crushing 19 atmospheres of pressure but the sensation is strangely comforting. Almost sedative. She continues her descent into the submarine night, a silky silence broken only by the rhythm of her heart slowly beating. Her wetsuit is now compressed paper-thin and the pressure in her eardrums builds to a mild pain that rapidly becomes intense and then quickly unbearable. She feels as if her ears are about to burst.

Looking down, she spots two more safety divers illuminating a small platform and she slows the descent imperceptibly, checking her hold on the sled as it finally hits its mark. A "man on the Moon" moment. She sees shimmers of reflected light flashing below her feet as the safety divers illuminate a digital depth gauge attached to the base structure. Its red lights blink 224 m. More than 70 m below the cruising depth of nuclear submarines! A nod, thumb, and forefinger touching okay and a smile to her safety divers, she reaches back over her shoulder and pulls a lanyard. A sudden rush of air is released as the cylinder tucked inside her wetsuit explodes compressed air into a lift-bag, transforming her from an aquatic mammal into a human missile headed for the surface.

Her face is calm as she streaks to the surface, her face shrouded in a cloud of bubbles. Black slowly becomes dark blue and dark blue slowly becomes the lighter shades of the shallow water before, finally, the silhouette of the platform comes into view. Faster, faster, she approaches the light at the surface. Special sensors in her brain detect an acidic content in her blood from the build-up of carbon dioxide and signal to her lungs to expel it. It is this carbon dioxide, and not lack of oxygen, triggering her urge to breathe.

As she closes in on the surface, she feels an intensifying pressure inside her chest, almost like an inflating balloon. *Breathe out!* her body is telling her. *Get rid of that carbon dioxide. Breathe out!* 20 m. *Hang on*, she tells herself, marshaling all her will to fight the expanding sensation in her thoracic cavity. *Don't panic. You've been here before.* She can feel herself being pulled inexorably to the surface by the billowing lift-bag. The surface can't come soon enough. Her heart thumps and her ears ring loudly with its thumping as she begins to experience the insidious and painful sensations of hypoxia. Her extremities gradually begin to burn like the distant pain of a dentist's drill penetrating the anesthetic. Lactic acid, another by-product of hypoxia, elevates the pain to almost intolerable levels. She summons her concentration as her consciousness begins to shrink from lack of oxygen, her lungs screaming for her to break the surface as she struggles to overcome every human instinct to simply breathe.

She concentrates on expelling air as she looks up, the shimmering undersides of the swells quickly becoming more distinct, their crests visible as thin bubbly streaks as she closes in on the surface. With the surface in sight, two freedivers join her for the final journey to the surface, kicking gracefully with their long fins. Like a sub-aquatic projectile, the new No Limits world record-holder slices upward, pushing her head upright like a seal pushing its nose through a hole in the ice. Finally, the water lightens and she explodes onto the surface in a flurry of bubbles, her upper body shooting above the water before tumbling back down like a surfacing dolphin. The wind and waves and bright tropical sunlight slap suddenly against her face as she glances at her watch. It shows an elapsed time of 3 min and 29 sec. She senses the invigorating sound of the surface and gulps a lungful of air, another, and finally a thumbs-up and a smile.

HOW DEEP CAN YOU DIVE?

What you have just read is a fictional account of a female freediver performing a world record in the No Limits category (Panel 1.1) – a mark currently held by Tanya Streeter (Figure 1.1), a Cayman Islands native who plunged to 160 m in 2002 (the male record is held by Austrian, Herbert Nitsch, who dived to 214 m in 2007).

The depths to which Streeter and Nitsch descend were not thought humanly possible until very recently. As late as the 1960s, diving physiologists and doctors were convinced the immense water pressure below 100 m would crush the human chest cavity. After legendary freediver, Jacques Mayol, and other pioneers of the sport extended the human deep-diving envelope and descended below this depth, scientists were forced to re-evaluate their predictions and discovered that exposure to the pressures of deep water causes the body to respond in unexpected ways. The lungs contract to the size of an orange, the blood flowing to the extremities is re-routed to the vital organs, and the heart rate slows to fewer than 10 beats per minute. This series of responses comprise the *mammalian diving reflex* (MDR), a metabolic switch recognized by researchers as a remnant of our aquatic origins. Since the late 1960s, scientists have studied, tested, and probed some of the best freedivers in the

Panel 1.1. No Limits freediving

No Limits is one of the five freediving depth disciplines:

1. Constant Weight, in which the athlete dives to depth following a guide line. "Constant" refers to the fact that the athlete is not allowed to drop weights during the dive.
2. Constant Weight Without Fins follows identical rules to Constant Weight, except no fins are allowed.
3. Free Immersion is a discipline in which the freediver uses a vertical guide line to pull him/herself down to depth and back to the surface. Again, the athlete is not allowed to release weights.
4. Variable Weight uses a weighted sled for descent and the freediver returns to the surface by pulling themselves up along a line or swimming while using fins.
5. No Limits is the freediving discipline that permits the athlete to use any means of breath-hold diving to depth and return to the surface as long as a guide line is used to measure the distance. Most freedivers use a weighted sled to dive down and use an inflatable bag to return to the surface. It is the discipline immortalized in Luc Besson's film *The Big Blue* and thanks to a spate of recent deaths, it has sparked the interest of newspapers, which inevitably classify freediving as an "extreme" sport.

world in an attempt not only to discover how this mechanism helps freedivers perform such deep dives, but also to learn exactly what the body is capable of achieving. After several dozen freediving studies, scientists have discovered that with training, the MDR response can be amplified. However, this research has not generated the data necessary to answer the question the media, freedivers, and diving physiologists have been asking since freediving became a sport over 50 years ago: How deep can a diver descend on one breath of air? To answer this question and to understand the problems faced by future No Limits freedivers, we must turn to some of the world's leading diving scientists.

No Limits research

Dr. Erika Schagatay is physiologist at Lund University, Sweden, who has been investigating the responses of freedivers for several years. A professor of animal physiology and an active diver, Dr. Schagatay (Figure 1.2) is also a freediving instructor. Her interest in freediving physiology began when she met native freedivers who performed far better than was suggested in the medical literature at the time. Since 1988, she has studied the physiology of several diving tribes, including the Japanese Ama and the Indonesian Suku Laut and Bajau. More recently, she has focused her research on competitive freedivers.

Figure 1.1. This photo shows No Limits world record-holder, Tanya Streeter, performing a No Limits dive. In 2002, Streeter, a Cayman Islands native, broke the men's No Limits record by diving to 160 m near the Turks and Caicos Islands, a record that was broken later that year by French diver, Loïc Leferme (whose record was broken by Herbert Nitsch). As the photograph of Streeter shows, the No Limits discipline is performed with the diver positioned in the vertical position and uses the same principle as Statti employed in what could loosely be considered the first "No Limits" dive many years ago. The reason this position confers an advantage upon the diver is based upon simple physics and physiology. When descending in the vertical position, the hydrostatic pressure affects the lower parts of the body first, but as the body is compressed, a translocation of blood from the lower to the upper extremities occurs – a mechanism that is more efficient if the body is in the head-up position. Using this position, No Limits divers ride a sled to the bottom of a cable, at the end of which they detach themselves from the sled and pull a pin, releasing compressed air from either a cylinder placed inside their suits or into a lift-bag placed on the sled structure. In a cloud of bubbles, they then ascend to the surface. Courtesy Tanya Streeter.

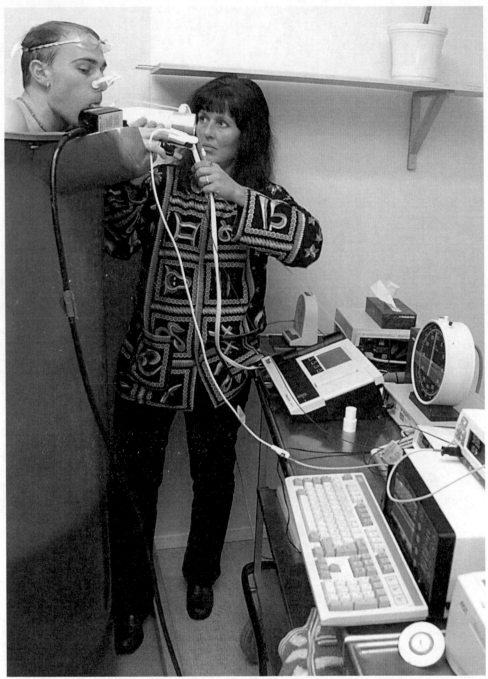

Figure 1.2. Erika Schagatay, a professor at Mid Sweden University, is one of the world's leading researchers on the subject of freediving physiology. Courtesy Erika Schagatay.

Much of Dr. Schagatay's research has been directed at determining the physiological criteria comprising a record-breaking freediver and to find answers to one of the most obvious questions: Can the body adapt to increasing pressure and hypoxia through training or are the most successful freedivers born and not made? As a part of one of her studies, she has measured the strength of the MDR (Panel 1.2) in trained and untrained divers using the drop in heart rate as an index [1]. For example, when divers of the Swedish National Freediving Team submerged their face and held their breath, their heart rates dropped by half, but by manipulating the temperature of the water, Dr. Schagatay was able to elucidate additional responses. Lowering the temperature resulted in an even stronger response, causing the heart rate to drop even more. In contrast, in the non-diving group, the reduction in heart rate was significantly smaller, the average drop being only between 20 and 30%.

Panel 1.2. Mammalian diving reflex

The mammalian diving reflex optimizes respiration, which permits mammals to stay underwater for long periods of time. It is exhibited in aquatic mammals such as seals and otters but also exists in humans. The reflex is triggered by cold water contacting the face, which initiates the following changes:

1. Reduction in heart rate by 10–25%.
2. Blood circulation to the extremities is closed off – first the toes and fingers, followed by the hands and feet, and finally the arms and legs.
3. During deep dives, a blood shift occurs. This mechanism allows plasma to pass freely throughout the thoracic cavity, so the pressure remains constant.

Developing her studies further, Dr. Schagatay decided to compare the responses of the trained divers with two groups with a long history of breath-hold diving. First, she studied the Japanese Ama, divers who use various freediving techniques to harvest shellfish and seaweed – something they have been doing for more than 2,000 years. She also studied the Sea People, a tribe of Indonesians who live a semi-aquatic existence, spending up to 10 hr a day in the water. Surprisingly, there were no distinct differences between the diving responses of the Swedish Freediving Team and those of the Ama and Sea People. Elaborating on these studies, Dr. Schagatay decided to examine the non-divers again to see if they were able to extend their breath-holding time. After 2 weeks of freediving instruction, the non-trained divers were able to significantly extend their breath-hold time, which clearly suggested to Dr. Schagatay that elite divers do not possess any special gene that enables them to do what they do. Supporting the findings of these studies, Dr. Schagatay has published the results of investigations showing an improvement in breath-hold time by as much as 50% after just five successive attempts in a single day [2]. Novice freedivers attending introductory freediving courses experience similar increases in breath-holding ability, often nearly doubling their time in the course of a weekend.

While the results of these studies were remarkable, there were no data explaining *how* such increases were achieved, but Dr. Schagatay thought one of the least understood organs of the human body might provide some of the answers.

The spleen lies directly beneath the diaphragm, behind and to the left of the stomach, and is covered by peritoneum. Weighing about 200 g, it is purplish in color and varies in size in different individuals, but is usually about 120 mm long, 70 mm wide, and 25 mm thick. Its primary function is to destroy red blood cells – a task it completes together with the liver, but it is the organ's secondary function that was of interest to Dr. Schagatay. Because of the huge volumes of blood that circulate through it, the spleen serves as a blood reservoir that plays an integral part in the human diving response and may help freedivers extend their breath-holds and time at depth. The theory was a good one, since the splenic reservoir function is observed in many animal species (Weddell Seals, for example, are able to store 24 l of blood in their spleens) and according to the results of Dr. Schagatay's studies [2], it is one of many physiological adaptations freedivers can develop enabling them to dive deeper. Very simply, the spleen *shrinks* while diving, causing a release of extra blood cells.

Studies similar to Dr. Schagaty's have corroborated these findings. According to William E. Hurford, MD, writing in *The Journal of Applied Physiology* [3], the spleens of the Japanese Ama divers decreased in size by 20% when they performed dives to depths of between 20 and 30 m. At the same time, their hemoglobin concentration increased by 10%. Of interest to Dr. Schagatay was the finding that in some studies, untrained subjects demonstrated a smaller contraction and a less pronounced increase in hematocrit and hemoglobin than the Ama, suggesting splenic contraction may be subject to a training effect.

The mechanisms triggering splenic contraction are included in the MDR outlined later in this chapter, and include a peripheral vasoconstriction[2] resulting in an increase in blood pressure that, in turn, causes a low heart rate. However, it is the effect and timeline of the mechanism that hold particular interest among scientists because research is a little vague in defining when splenic contraction occurs, although it has been demonstrated that for complete splenic contraction to occur, more than one apnea must be fully initiated. Once this has been performed, the contraction normally occurs within 30 sec and upon reaching the surface, the spleen returns to its normal size after about 10 min. Other studies argue that the spleen's contraction and subsequent release of red blood cells are not that immediate and may take up to a quarter-hour of sustained freediving. Some researchers state even this time period is insufficient, and suggest that for freedivers to experience the full effect of splenic contraction, they should perform at least 30 min of sustained diving.

Regardless of which study is correct, the effect has implications for those freedivers performing future No Limits record attempts. The mechanism may also explain why dives performed later in a competition/training session are often deeper and longer than those performed at the beginning of a session.

---

[2]   Vasoconstriction means constriction of the blood vessels, while peripheral blood vessels are those not in the core of the body such as the blood vessels of the skin.

Also of interest to researchers is whether splenic contraction constitutes a part of the MDR and if it is involved in the short-term training effect when performing repeated breath-holding – a skill comprising an important component of a No Limits dive. A recent study investigated this by performing a simple study involving 20 volunteers of whom 10 had previously been splenectomized (spleens removed for medical reasons) and 10 had healthy spleens. Each subject performed five maximal-duration breath-holds with face immersion in cold water (10°C) with each breath-hold separated by a 2-min interval. In subjects with spleens, the hematocrit and hemoglobin concentration increased during the apneas and returned to baseline within 10 min of the last apnea being performed. The researchers also observed a delay in the physiological breaking point of breath-hold in this group. In the splenectomized group, an increase in neither hemoglobin nor hematocrit was observed, nor was a delay in the breaking point of breath-hold recorded. These results suggest splenic contraction occurs in humans as part of the MDR and splenic emptying may prolong repeated apneas. As freedivers continue to dive deeper, and diving physiologists continue to scratch their heads in search of an explanation, an organ that rarely gets a mention might just be one of the keys to continued improvement.

THE DANGERS OF NO LIMITS

"It's like Russian Roulette; the deeper you go, the more bullets are in the chamber."
            Neal Pollock, Research Associate, Center for Hyperbaric Medicine and
                                    Environmental Physiology, Duke University

While providing enticing possibilities for further investigation, the mechanism of splenic contraction is just one of several processes that may explain how freedivers do what they do. Given the steady progression of No Limits, it is obvious that appreciable margins still exist for improving depth records, although an increasing number of scientists suspect freedivers are fast approaching the ultimate depth. One of the leading researchers investigating the mechanisms comprising the physiological redline of freediving is Dr. Claes Lundgren (Figure 1.3), Director of the Center for Research in Special Environments at the University of Buffalo Medical School.

One of the mechanisms Dr. Lundgren has focused particular attention upon is the *blood shift*. This phenomenon has been reported in freedivers performing dives to relatively shallow depths, but the effect is more pronounced the deeper the diver descends. In divers attempting future No Limits world records, it may even result in serious injury. As Dr. Lundgren points out in his research, although the blood shift is a feature of the MDR, there are a number of dangers when subjecting the body to such stresses repeatedly. One of these is the rupture of the pulmonary blood vessels, which, in turn, leads to bleeding into the alveoli and subsequent pulmonary edema, a mechanism that has occurred in freedivers performing dives as shallow as 25 m. These events occur during the descent when the lung volume becomes smaller than

Figure 1.3. Claes Lundgren is the Director of the Center for Research in Special Environments in the School of Medicine and Biomedical Sciences at Buffalo University. He is the world authority on the lung at depth. While most researchers are reluctant to put a figure on how deep freedivers can dive because they keep being proved wrong, Lundgren is willing to stick his neck out. In an interview with *New Scientist* magazine in March 2001, he was quoted as saying: "I think we are now at the limit. If they really knew what they were putting their bodies though, I don't think they would dive." In the meantime, divers like Herbert Nitsch are planning a dive to 300 m. Courtesy University of Buffalo.

the residual lung volume (RLV). To better understand this mechanism, it is useful to review some basic diving physiology, which will also help us grasp what the maximum depth limit might be. To start with, we know that the effect of increasing depth is to compress the gas in the lung and for lung volume to be reduced in accordance with Boyle's Law. At a certain depth, total lung capacity (TLC) will decrease until it equals the RLV, after which descending further may result in alveolar hemorrhage and pulmonary edema [4, 5]. It is this TLC:RLV ratio that is the *theoretical* maximal depth limit for a freediver, but since freedivers regularly exceed their depth limits, based on this ratio, there must be other factors implicated. One of these is the high negative intra-thoracic pressure that develops as the freediver descends below 30 m, causing the chest wall to approach its elastic limit and resulting in a liter of blood being drawn into the thorax. The effect of this translocation is for the pulmonary capillaries to distend into the alveolar spaces, replacing air and causing a decrease in residual volume. Although this series of events may extend the depth limit, there is a physiological price to be paid by the freediver, since this blood shift increases the likelihood of alveolar hemorrhage as a result of the increased pressure inside the capillaries [5]. The reason(s) why some divers experience bleeding at a specific depth while others do not is a mystery to respiratory physiologists, but the factors affecting the pulmonary blood–gas barrier and those that cause alveolar hemorrhage are well documented. One mechanism is the decrease in elasticity of the respiratory membrane that may be caused as a result of the extreme ventilation [6] practiced by freedivers (Panel 1.3).

Panel 1.3. Packing

Before attempting a deep dive, freedivers obviously want to get as much air into their lungs as possible. One way they do this is "lung packing", a special respiratory maneuver that physiologists call glossopharyngeal insufflation (GI), which involves adding air to the lungs on top of a full inspiration. "Packing" means freedivers start their dives with a large air volume in the lungs, which adds to the main oxygen store of the body and adds space for carbon dioxide storage. Divers also use the "packing" technique on dry land to improve the flexibility of the chest and stretchability of the diaphragm. With practice, some divers are able to "pack" large volumes and expand the chest significantly, giving them an odd barrel-chest appearance.

The second mechanism is an increase in blood pressure in the pulmonary capillaries as experienced during exposure to cold water and immersion. It is a combination of these two mechanisms that may cause an increase in pulmonary capillary pressure that ultimately results in ultrastructural changes in a disruption of the blood–gas barrier. These changes, in turn, result in the alveolar hemorrhage as a result of stress failure of the capillaries [7, 8].

Dr. Lundgren suggests it is possible that those divers whose lungs rupture at a shallower depth may have abnormalities in their vessels. One such abnormality is *telangiectases*. This condition, which may be congenital, is one in which the small arteries are expanded and weak connections exist between the small arteries and veins making them more susceptible to rupture. Although, to date, there have been no cases of severe damage inflicted upon a freediver's pulmonary system, an accumulation of blood in the alveoli is not a natural event and if it occurs repeatedly may lead to a condition known as *hemosiderosis*. Hemosiderosis results in a scarring of the lung tissue and subsequent dysfunction of the pulmonary system. Diving physiology suggests that at a certain depth, bleeding into the lungs could cause permanent damage and possibly even death. This type of injury is occasionally visible in competitors who, upon returning from a freedive, are observed coughing up foam tinged with blood. According to Dr. Lundgren, author of the definitive text on the lung [5], such a sign suggests these divers are experiencing an extreme redistribution of blood, which may place an unhealthy load upon the heart to a

Figure 1.4.   In June 2005, Belgian Patrick Musimu redefined human limits by becoming the first man to dive deeper than 200 m (he reached 209.6 m) on a single breath of air. With a lung capacity of 9 l (normal is about 5 l) and an ability to hold his breath for 7 min, Musimu is the human version of a seal. His dive made him a celebrity in the elite world of freediving. According to scientists, Musimu's dive was beyond the realms of physiology, prompting many to warn of dire consequences for those attempting to repeat Musimu's dive. The warnings fell on deaf ears, because 2 years later, Herbert Nitsch plunged to a depth of 214 m. Courtesy Patrick Musimu.

degree that may cause the heart muscle to become overly distended. In fact, it is this distension of the central circulation that may explain the unusual arrhythmias often observed in freedivers immediately following a No Limits dive (an element of the diving response noticeably absent in other diving mammals). While such arrhythmias may not pose a high risk in healthy divers performing regular freediving activities, in susceptible individuals who may have congenital cardiovascular defects and in divers attempting No Limits dives, it may lead to a fatal outcome (Figure 1.4).

Hemosiderosis isn't the only danger faced by wannabe No Limits freedivers. Those intending to plunge below 200 m on one breath of air also have to contend with dangers such as high pressure nervous syndrome (HPNS) and decompression sickness (DCS). Consider the case of German freediving champion, Benjamin Franz. In 2002, Franz returned from a series of 100-m training dives and was rushed to a hyperbaric facility for treatment for DCS. The cause of his accident, which resulted in him being confined to a wheelchair for nearly a year, was gas liberation into his tissues and cardiovascular system. For those like World No Limits Record-Holder, Herbert Nitsch, who is planning a 300-m dive, there is another potential physiological showstopper: HPNS. HPNS is a group of abnormal behavior and central nervous system (CNS) changes occurring during dives deeper than 150 m [9]. The syndrome includes the following phenomena: tremors, headache, changes in electroencephalogram (EEG) activity, reduction in psychomotor and cognitive functioning, nausea, fatigue, and insomnia [10]. As Dr. Lundgren points out, the nature of HPNS is a complex one and individuals vary dramatically in susceptibility, but it is known that the faster the rate of descent, the more pronounced the symptoms [5, 11]. Since a typical No Limits attempt requires freedivers to descend at a rate approaching 3 m/sec, this syndrome presents a serious risk.

Given the problems faced by those considering a No Limits dive, it would seem sensible to ask which elements of training might be manipulated to ensure the greatest chance of success. Researchers agree that current freediving training is specific and effective, since it is known to improve the MDR. Unfortunately, for freedivers with record-setting aspirations, those same scientists also agree there is no method of training that can reliably minimize the dangers outlined above. The reason, Dr. Lundgren explains, is that many of these risks are defined as injury mechanisms and not physiological adaptations. By performing repeated exposures to depth, Dr. Lundgren argues, freedivers may continue to improve the MDR, but a point will be reached at which this performance benefit will be outweighed by the risk of sustaining cumulative damage as a result of repeated exposures to these injury-inducing mechanisms.

IN PURSUIT OF THE ULTIMATE DEPTH

> "Somebody will eventually find out what the absolute limit is. And the way that person is going to find out is by not coming up alive."
> Dr. Claes Lundgren on the subject of the limits to No Limits freediving

When conventional methods of training have been exhausted and the records appear to be insurmountable, freedivers may turn to science to help them dive deeper (Figure 1.5). Since many of the problems encountered during freediving relate to the compressibility of gases, one solution may be to use a breathing mixture that is incompressible. The obvious danger would be drowning and drowning mammals usually inhale water, which can cause tissue damage and, ultimately, death. However, if the lung is filled with a solution that does not mix with water, then no such damage should occur. Medical technology has designed artificial kidneys and artificial lungs in which the blood of the patient flows outside the body and exchanges gases with the external environment. It would be reasonable to expect that a gas exchanger could be designed using artificial breathing membranes enabling a freediver with liquid-filled lungs to obtain the necessary oxygen from the sea water.

Figure 1.5.   Herbert Nitsch performing a dive in the No Limits category. An Austrian freediver who has held world records in eight freediving disciplines, Nitsch is the deepest man on Earth following his world record in the No Limits discipline. A part-time pilot for Tyrolean Airways, Nitsch set the world record in Spetses, Greece, in June 2007, when he descended to 214 m, beating his own record of 183 m set the previous year. Courtesy Herbert Nitsch.

Whether using such techniques could still be considered freediving is open to debate but it is interesting to explore the feasibility.

Many scientists have noticed the functional similarity between fish gills and mammalian lungs and have wondered whether a human version of such a lung could be bioengineered to breathe water, assuming sufficient oxygen was present. Upon closer examination of this intriguing possibility, however, even the most optimistic scientist acknowledges that from a structural perspective, there are significant differences. In the case of the fish gill, the gas exchange is between water and blood, the structure of the gill consisting of a number of parallel planes in which gases are exchanged between water and blood by diffusion. From a bioengineering perspective, the system has the appearance of a radiator in which water flows directly past several capillaries that exist in each gill unit. It is this difference in geometry that presents the biggest problem to those scientists hoping to create a homo aquaticus, since the lamellar structure of the fish gills acts very differently from the spherical (alveoli) gas exchange units that humans must rely upon. The reason is simple physics. A hypothetical fish bioengineered with spherical gas units such as ours would not survive because gases diffuse much more slowly in a sphere than they do in a lamellar structure. If the opposite were to happen and water was introduced into a human lung, then the spherical shape of the alveoli would result in an increased diffusion time and would therefore adversely affect carbon dioxide elimination. The bioengineering solution to this problem is surprisingly simple in theory, although a little more difficult in practice, since it would involve cutting a hole in the bottom of the lung. This may sound simple but, in practice, the lung is not a single-space bag so this bioengineering feat would require that all the millions of alveoli be attached to a drainage plumbing system that could release water from the system. If this were achieved, it would enable exhalation to be performed through the base of the lung – a situation in which the lung could probably expel sufficient carbon dioxide. By performing this rather elaborate surgery, the scientists would create a situation in which solution is pumped through the lungs instead of pumping into and out of the same section of lung. This would, in effect, create a similar situation that exists in a fish gill, since there would be a continuous fresh flow of water through the respiratory structure. Surgical intervention as described is probably several years in the future, but the concept is currently being investigated for scuba-diving applications by Israeli inventor, Alan Bodner, whom we meet in Chapter 9.

In the meantime, divers such as Herbert Nitsch will continue to confound the experts by performing deeper and deeper dives driven by human nature that inspires such individuals to explore the unknown and achieve what is considered impossible. Unfortunately, for those divers who choose to challenge the depths and push the conventional limits, today's current level of understanding of the way the body functions at depth does not allow for informed strategies to be formed or for estimates to be made regarding the physiological limits of this sport. In the scuba-diving community, the diver who dives below the recommended depth of 40 m is sometimes seen as a risk-taker and even dangerous, but freedivers who extend the limits of their sport are thought daring and even worshipped. A lack of knowledge is not a problem confined solely to the sport of freediving, however. In 1980, Reinhold Messner ascended to the summit of Everest without using supplemental oxygen, despite the dire warnings of altitude physiologists who insisted such a feat would be impossible. Since then, there have been numerous repeat ascents of not only Everest, but also other peaks above 8,000 m. Thirty years later, the medical problems encountered by extreme altitude climbers are incomplete but this does not deter the mountaineers. Regardless of the opinions of physiologists or physicians, No Limits freedivers will continue to pursue their activity to the fullest extent.

REFERENCES

[1]   Schagatay, E.; Andersson, J. Diving Response and Apneic Time in Humans. *Undersea & Hyperbaric Medicine*, **25**(1), 1–9 (1998).

[2]   Schagatay, E.; Andersson, J.P.; Hallen, M.; Pålsson, B. Selected Contribution: Role of Spleen Emptying in Prolonged Apneas in Humans. *Journal of Applied Physiology*, **90**(4), 1623–1629 (2001).

[3]   Hurford, W.E.; Hong, S.K.; Park, Y.S.; Ahn, D.W.; Shiraki, K.; Mohri, M.; Zapol, W.M. Splenic Contraction during Breath-Hold Diving in the Korean Ama. *Journal of Applied Physiology*, **69**(3), 932–936 (1990).

[4]   Hong, S.-K. Breath-Hold Diving. In: A.A. Bove (ed.), *Bove and Davis' Diving Medicine*, 3rd edn, pp. 65–74. W.B. Saunders Company, Philadelphia, PA (1997).

[5]   Lundgren, C.E.G.; Miller, J.N. (eds). *The Lung at Depth*. Dekker, New York (April, 1999).

[6]   Dreyfuss, D.; Basset, G.; Soler, P.; Saumon, G. Intermittent Positive-Pressure Hyperventilation with High Inflation Pressures Produces Pulmonary Microvascular Injury in Rats. *American Review of Respiratory Disease*, **132**, 880–884 (1985).

[7]   Kiyan, E.; Aktas, S.; Toklu, A.S.; Hemoptysis Provoked by Voluntary Diaphragmatic Contractions in Breath-Hold Divers. *Chest*, **120**, 2098–2100 (2001).

[8]   West, J.B.; Mathieu-Costello, O. Structure, Strength and Failure of the Pulmonary Blood Gas Barrier. In: J. Milic-Emili (ed.), *European Respiratory Monograph: Respiratory Mechanics*, pp. 171–202. ERS Journals Ltd, Sheffield, UK (1999).

[9]   Jain, K. High Pressure Nervous Syndrome (HPNS). *Acta Neurologica Scandinavia*, **90**, 45–50 (1994).

[10] Seo, Y.; Matsumoto, K.; Park, Y.M.; Mohri, M.; Matsuoka, S.; Park, K.P. Changes in Sleep Patterns during HeO2 Saturation Dives. *Psychiatry Clinical Neuroscience*, **52**, 141–142 (1998).

[11] Bennett, P.B.; McLeod, M. Probing the Limits of Human Deep Diving. *Philosophical Transactions of the Royal Society London*, **B-304**, 105–117 (1984).

2

Technical and Saturation Diving

In the world of recreational scuba-diving, you can earn your Deep Diver Specialty Course by paying US$175 and completing two open-water dives down to depths of between 18 and 40 m. For some divers, however, the 40-m limit prescribed by PADI (Professional Association of Diving Instructors) is not deep enough. Fortunately, for those wishing to venture deeper, there is the world of technical diving.

Technical diving[1] is usually defined as diving that includes dives deeper than 40 m, required stage decompression, diving in an overhead environment beyond 40 linear meters from the surface, accelerated stage decompression and/or the use of multiple gas mixtures in a single dive.

Although commercial divers venture deeper, technical divers fall into a special category, since, by utilizing open-circuit equipment, they face infinitely greater risks. In fact, as we shall see in this chapter, it is no exaggeration to say that this elite group of divers work on the ragged edge of technological and physiological knowledge.

The deepest dive achieved by a technical diver using open-circuit scuba is 330 m – a mark set by French diver, Pascal Bernabé, on July 5th, 2005. However, deep as this may be, there is another group of divers who regularly dive even deeper. Saturation (SAT) divers operate at extreme depths as deep as 500 m, breathing exotic cocktails of helium, oxygen, and hydrogen. "Saturation" refers to the fact that the diver's tissues have absorbed the maximum partial pressure of gas possible for that depth due to the diver being exposed to breathing gas at that pressure for prolonged periods.

Here, in Chapter 2, we consider the world of extreme mixed-gas diving. We discuss how revolutionary technologies such as rebreathers will allow technical divers to continue to dive ever deeper and how divers may one day overcome physiological problems such as decompression sickness (DCS) by simply popping a pill.

---

[1] The term "technical diving" was first coined in 1991 by Michael Menduno, editor of *AquaCorps*.

TECHNICAL DIVING

Extreme technical diving requires extraordinarily high levels of training, experience, fitness, and logistical support. As of 2009, only eight technical divers are known to have ever dived below 240 m using open-circuit scuba equipment. More people have walked on the Moon and those who understand the perils of this high-risk underwater activity would argue that traveling to the Moon is less dangerous! The deep-diving daredevils of the technical diving community push far into the dark labyrinths of extreme ocean depths. During these extreme excursions, they encounter myriad dangers ranging from disorientation and oxygen toxicity to high pressure nervous syndrome (HPNS) and nitrogen narcosis. Some succumb to the dangers of breathing helium, a gas that can reduce even the most prepared diver to a nervous, quivering wreck. If they survive the gauntlet of these hazards, there are the problems of ascent. If they ascend too quickly, all the nitrogen and helium that has been forced into their tissues under pressure can fizz into tiny bubbles, causing a condition known as the "bends", which may result in paralysis and death.

If you ask recreational scuba-divers about technical diving, you will get different answers. While newly minted scuba-divers may talk about their 20-m dive for months, experienced technical divers frequently plan dives in the 60–100-m depth range. However, there is an elite cadre of technical divers that plan even deeper dives. In the same way as the military seductively draws in new recruits through the imagery of high technology and personal challenge, sport divers are enticed to experience the "new frontiers" afforded by technical diving, but apart from the depth that technical divers reach, there is some disagreement about what the term "technical diving" means. Some divers argue that technical diving is any type of scuba that is considered a higher risk than conventional recreational diving, while others seek to define technical diving by reference to the use of decompression. A minority in the diving community contend that certain non-specific higher-risk factors require diving to be classed as technical diving. PADI, the largest recreational diver training agency in North America, has adopted this as their definition of technical diving:

> "Diving other than conventional commercial or recreational diving that takes divers beyond recreational diving limits. It is further defined as an activity that includes one or more of the following: diving beyond 40 meters, requiring stage decompression, diving in an overhead environment beyond 40 linear meters from the surface, accelerated stage decompression and/or the use of multiple gas mixtures in a single dive."

PADI's depth-based definition is derived from the fact that breathing regular air at pressure causes a progressively increasing amount of impairment due to nitrogen narcosis that may become serious at depths of 30 m or greater. Increasing pressure at depth also increases the risk of oxygen toxicity based on the partial pressure of oxygen in the breathing mixture. For this reason, technical diving often includes the use of breathing mixtures other than air. PADI's mention of decompression alludes

Figure 2.1. Technical diving involves diving beyond the normal limits of recreational diving – a feature reflected in the amount of equipment technical divers need, as shown in this photo. Courtesy Wikimedia.

to the fact that technical dives may be defined as dives in which the diver cannot safely ascend directly to the surface due to a mandatory decompression stop. This type of diving obviously implies a much larger reliance on specialized equipment (Figure 2.1) and training, since the diver must stay underwater until they have completed their decompression stop(s).

It is this reliance upon specialized equipment, combined with the necessity for specialized training, that increases the requirements for risk acceptance, preparation, and level of danger. Additionally, a defining feature that sets technical diving apart from its recreational counterpart is that there is a considerably greater risk and danger from DCS, drowning, oxygen toxicity, nitrogen narcosis, and equipment failure. These risks stem from a number of factors primarily related to the depth, duration, and restrictions of the dives that technical divers perform. A classic example is decompression. Technical diving almost inevitably involves decompression diving, which, in itself, introduces a significant risk.

Decompression sickness

As soon as a diver submerges beneath the water, he/she begins to incur a decompression debt. The debt is created by a greater amount of gases dissolving in the blood and cells as the greater depth causes an increase in the partial pressure relative to the conditions at the surface, where the pressure is 1 atmosphere. When the diver begins his/her ascent, these gases leave solution and may cause life-threatening bubbles in the bloodstream and tissues. The amount of dissolved gas in the blood and other tissues is a factor of time and pressure. The dissolution of gas is not instantaneous, but happens over time until equilibrium is reached. This means that even a shallow dive to 10 m for a long period of time can cause the dissolution of a substantial amount of gas as a deep dive for a short period of time. The limits to deep diving are set by factors related to the increasing mass of the water column above the diver as the diver descends. Decompression problems occur during the ascent phase of a dive. The diluent gas in the divers' breathing mixture dissolves in the tissues in proportion to the solubility of the gas in tissues and the partial pressure of the diluent gas. During the ascent, the partial pressure of the diluent gas decreases, which causes a decreasing volume of gas to remain in solution in the tissues. The rate at which this occurs is determined by the rate of ascent and is perhaps the most important aspect of any dive. To ensure that the excess diluent gas can be eliminated via diffusion across the epidermal tissue, lungs, or other mucous membrane body surfaces, with the residual gas remaining in solution, the diver must ascend at a prescribed rate. If the ascent rate is exceeded, gas bubbles may form in the diver. When the gas bubbles are large and numerous, or located in particularly vulnerable tissues such as the spinal cord and joints, they may cause a painful and potentially seriously debilitating condition known as DCS. A diver at the end of a long or deep dive may need to perform decompression stops (Figure 2.2) to avoid DCS. This is because metabolically inert gases in the diver's breathing gas, such as helium, are absorbed into body tissues when breathed under high pressure during the deep phase of the dive. These dissolved gases must slowly be released from body tissues by stopping at various depths during the ascent to the surface. In the last decade, most technical divers have favored deep first decompression stops because research suggests this will reduce the risk of bubble formation before the long shallow stops.

Compounding the problem of decompression is the challenge posed by breathing gas mixtures. To reach the depths attained during technical diving, divers must use exotic gas blends. This is because breathing a mixture with the same oxygen concentration as air (roughly 21%) at depths greater than 55 m results in a rapidly increasing risk of severe symptoms of oxygen toxicity – a syndrome that may prove deadly. The first sign of oxygen toxicity is often a convulsion without warning. Occasionally, the diver may experience warning symptoms prior to the convulsion. These symptoms may include visual and auditory hallucinations, nausea, twitching, irritability, and dizziness. More often than not, a convulsion at depth is usually fatal because the regulator will fall out and the diver will drown.

Increasing depth also causes nitrogen to become narcotic, resulting in a reduced ability to react or think clearly – a syndrome known as *nitrogen narcosis*. By adding

Figure 2.2. A technical diver performing a decompression stop. A "deco" stop is a period of time a diver must spend at constant depth following a dive to eliminate absorbed inert gases from the body to avoid decompression sickness. A technical diver at the end of a long or deep dive may need to perform multiple "deco" stops at various depths during the ascent to the surface. In recent years, technical divers have increased the depth of the first stops, to reduce the risk of bubble formation before long shallow stops. Courtesy Mark Ellyat.

helium to the breathing mix, divers can reduce these effects, as helium does not have the same narcotic properties at depth. These gas blends can also lower the level of oxygen in the mix, thereby reducing the danger of oxygen toxicity. Once the oxygen fraction is reduced to below 18%, the mix is known as a hypoxic mix, since it does not contain sufficient oxygen to be used safely at the surface. Another technique employed by technical divers to reduce the risk of nitrogen narcosis is to breathe enriched oxygen breathing gas mixtures during the beginning and ending portion of the dive. For example, while at maximum depth, it is common to use a gas known as "trimix", which adds a fraction of helium to and replaces nitrogen in the diver's breathing mixture. Another tactic designed to reduce the time required to rid themselves of most of remaining excess inert gas in their body tissues and to reduce DCS risk is to use pure oxygen during the shallow decompression stops.

One gas that technical divers use to avoid nitrogen narcosis is heliox, a mixture of helium and oxygen. It is used as a breathing gas only for extreme diving depths and the exact gas fractions in the mixture are determined by the intended depth. For example, imagine a diver making a dive to 100 m. The pressure at 100 m is 11 atmospheres (pressure increases at a rate of 1 atmosphere for every 10 m of depth, so at 100 m, the pressure exerted upon the diver is the atmospheric pressure *plus* the pressure of 10 atmospheres), or 11 *bar*, to use the diving vernacular. Because the diver knows that oxygen toxicity occurs at a partial pressure of 1.6 bar, he/she must choose a gas blend that will have a lower oxygen partial pressure than 1.6 bar. To err on the side of caution, he/she might choose a mixture comprising 14.5% oxygen and 85.5% helium. This oxygen fraction would minimize the chance of oxygen toxicity, since the partial pressure at 100 m would be 1.595 bar ($pO_2 = 0.145 \times 11 = 1.595$ bar). While heliox is an easy blend to manipulate, perhaps its greatest attraction to technical divers is that it does not contain any nitrogen, so the risk of DCS is eliminated. However, one shortcoming of heliox is that helium is a highly efficient conductor of heat so a diver breathing a helium blend will chill far more quickly than an air-breathing diver. But, of course, an air-breathing diver is limited to dives of only 40 m! Unfortunately, while heliox is a favorite gas blend among technical divers, the gas is expensive and, at present, most of the world's helium reserves are in Texas and those reserves are running low. These circumstances have forced technical divers to consider other blends such as hydrox. If any blend can be considered the perfect mix for technical diving, it may be hydrox, which is a blend of hydrogen and oxygen. Although the mixture is still in the experimental stages, it has produced good results, since it is easy to breathe, incurs no DCS risk, and the gas conducts heat slowly, unlike helium. Its only real flaw comes from the extremely flammable nature of hydrogen – a stigma that has followed the gas since the *Hindenburg* disaster.

Another popular gas mix that extends depth considerably is trimix, a blend comprising oxygen, nitrogen, and helium. To create trimix, the helium is added as a diluent for the nitrogen content and the oxygen level is also reduced, depending on the depth desired. With the reduced levels of nitrogen and oxygen, the maximum safe depth can exceed 100 m. The downside to trimix is that unless the oxygen content is kept between 18 and 21%, a travel gas[2] is usually required for descending to the safe breathing zone. Inevitably, this means extra cylinder requirements and gas planning, but this is all part and parcel of technical diving so is considered a minor inconvenience.

So far, we have discussed the risks of technical diving and the gas mixes required. Before we can take a step into the future of this cutting-edge sport, we must now consider the equipment. Technical divers use a truly extraordinary amount of equipment (Figure 2.3). In fact, witnessing a diver gearing up for a deep mixed-gas dive is akin to watching an astronaut prepare for a spacewalk. In addition to the standard drysuit, the modern-day technical diver is encumbered by two or three of everything, highlighting a (necessary) compulsion for redundancy that further underlines the dangerous nature of the activity. As you can see in Figure 2.3,

---

[2]   A travel gas is simply one that is breathed until the target depth is achieved, whereupon a gas switch is performed and the diver begins breathing the trimix.

Figure 2.3. Here, you get some idea of the set-up required for technical diving. The tech-rig pictured is a good illustration of the redundancy required to safely perform technical dives and also the amount of gas that needs to be carried to perform all the decompression stops. Courtesy Mark Ellyat.

technical divers usually carry at least two cylinders, each with its own regulator. In the event of a failure, the second cylinder and regulator act as a backup system. Technical divers therefore increase their supply of breathing gas by connecting multiple high-capacity diving cylinders. The technical diver may also carry additional cylinders, known as *stage bottles*, to ensure adequate breathing gas supply for decompression with a reserve for bail-out in case of failure of their primary breathing gas. The amount of equipment worn by the diver in Figure 2.3 is typical for a technical diver and while it may appear unwieldy, the set-up is fairly standard.

EXTREME DIVING

Once a diver has chosen his gas mix, decided which equipment configuration he/she is going to use, and planned the dive, all that remains is the dive itself. Perhaps the best way to explain what happens during a technical dive is to follow the exploits of an extreme technical diver during one of the deepest dives ever recorded. What follows is an account of Mark Ellyat's then world-record dive to 313 m in December 2003.

Figure 2.4. In 2003, Mark Ellyat broke the record for the world's deepest dive using scuba equipment, reaching 313 m off the coast of Phuket, Thailand, following a dive lasting 7 hr. For those interested in exploring the extreme side of technical diving, Mark has published a book – *Ocean Gladiator* – that describes his exploits deep beneath the ocean. Courtesy Mark Ellyat.

One thousand feet (305 m) was long regarded as the 4-min mile of open-circuit scuba-diving and still marks the gold standard for those in the technical diving community. Those attempting depth records using open-circuit scuba usually do so in fresh-water sink holes, since these environments provide reasonably predictable conditions, which makes the staging of cylinders and equipment a little easier. However, even when diving in a sink hole, the technical diver still has to consider huge differences in surface and bottom temperatures and unpredictable currents. The record Ellyat (Figure 2.4) was attempting to break was set by John Bennett in November 2001. Bennett was no stranger to the extraordinary demands of extreme deep diving. In June 2000, he had established a world depth record of 254 m in the warm waters off Puerto Galera in the Philippines. However, setting a world record was not enough for Bennett, who immediately began planning an even deeper dive that would break the thousand-foot barrier. He achieved his goal by plunging to 308 m in November 2001. In March 2004, Bennett went missing in South Korea in just 45 m of water and was later declared dead.

Like Bennett, Ellyat's goal was a world-record dive, but it almost did not happen. In February 2003, the diver with almost 3,000 dives under his weight belt almost

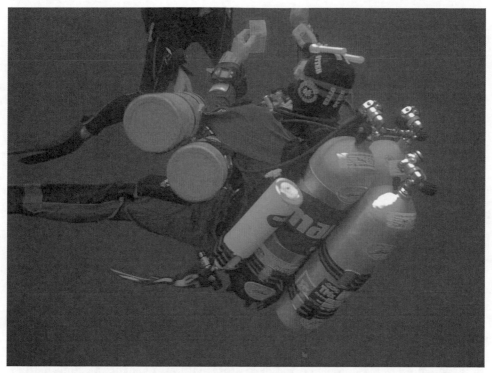

Figure 2.5. Mark Ellyat kitted out in the equipment he used to break the record for the world's deepest dive. Courtesy Mark Ellyat.

made his last dive during a practice dive to 260 m; Ellyat's decompression schedule proved inadequate and he sustained serious injuries. The doctor advised him never to dive again, which, to any extreme sportsman, is akin to waving the proverbial red rag to a bull. Ellyat ignored the medical advice and started planning his attempt on Bennett's record.

Embarking upon such a venture was something akin to planning a space mission, requiring months of planning just to configure the equipment, the complexity of which was formidable. The first problem faced by any diver attempting to dive so deep is the sheer number of cylinders required (Figure 2.5) to carry the huge amounts of gas required for what may be a dive lasting 10 hr or more. Whereas a recreational diver only has to worry about a single cylinder on his/her back, for the Mark Ellyats of the diving world, cylinders must be brought along for the descent, the bottom (for decompression), drysuit inflation, and decompression stops during the ascent. For this record attempt, Ellyat wore an Otter "Velvet-Skin" purple-colored drysuit, constructed of a special membrane material designed to be strong but flexible. Because he needed to carry multiple steel cylinders on his back, Ellyat had to use special dual bladder wings to provide sufficient buoyancy.

Another key part of any successful dive is the support team. For his record-breaking dive, Ellyat employed the services of 14 experienced deep divers in addition to a paramedic for any medical contingencies. In the week leading to the dive, the support team busied themselves with briefings, discussions of dive profiles, potential gas problems, abort scenarios, and setting the depths for the support divers. Once the dive profile had been approved, some team members went about assembling the dive platform and fine-tuning contingency evacuation plans while others started the complex process of mixing Ellyat's gas blends.

Ellyat's dive site was on the edge of the continental shelf at 450-m depth, 60 km offshore from Phuket, Thailand – a site chosen due to the high quality of medical support and availability of military hyperbaric chambers. If Ellyat had a problem, he knew he could be at a hyperbaric facility quickly thanks to a 600-horsepower engine bolted to the back of one of the boats. The descent proceeded smoothly until 250 m, when Ellyat noticed minor HPNS symptoms, although he was not sure whether the shakes were helium-induced or the result of the icy water. On reaching 300 m, he noticed the contents gauge showed the turn-around pressure but he continued on down to 313 m. The descent had taken just 12 min. At the bottom of the dive, the water temperature was just 3°C. Spending no more than 60 sec there, he collected a marker to verify his record-breaking depth and began his ascent. In common with Bennett, Ellyat's ascent profile was based on intuition and knowledge of the emerging science of deep decompression that suggests decompression benefits can be gained by stopping deep for short periods of 30 sec or less. However, too many stops below 200 m may add to the overall decompression penalty. The problem is no one really knows for sure. Ellyat's ascent rate on leaving the bottom was a conservative 18 m per minute. He completed the deep stops without incident, meeting his first support diver at 90 m, where he was handed a cylinder of trimix. After completing his 90-m decompression stop, Ellyat continued on to complete more stops at 75 and 60 m. The ascent took 6 hr and 36 min, during which Ellyat used 24 cylinders brought down to him by support divers.

Ellyat's descent to 313 m was an epic dive, and one that blurred the distinction between commercial diving and the sport diving community. As the description of Ellyat's dive illustrates, venturing into the realm of decompression and mixed gas requires far more training, equipment, and knowledge of operational disciplines than most technical divers have or are willing to acquire. Furthermore, diving to extreme depths using cumbersome cylinder rigs and multiple gas mixes is a dangerous pursuit, as evidenced by the death of John Bennett and several other divers in the technical diving community. There has to be a better and safer way.

REBREATHERS

The equipment that may represent the future of technical diving is the rebreather (Figure 2.6), a breathing system enabling the diver to retain and reuse some or all of the expired gas. The problem with open-circuit (so-called because once the gas leaves the diver, it is no longer part of the breathing cycle) scuba is that the diver typically only uses about a quarter of the oxygen in the air that is breathed in. The rest is exhaled along with nitrogen and carbon dioxide. It is a very wasteful and inefficient

Figure 2.6. The Evolution is typical of the current generation of closed-circuit rebreathers. Compact, streamlined, and weighing only 24 kg, diving with this CCR is a dream compared with the cumbersome tech-rigs that are usually associated with technical diving. Photo John Bantin. Courtesy Ambient Pressure Diving.

system and as depth increases, the inefficiency of the open-circuit system is compounded, since, because of the increased pressure, even more gas is lost with each exhaled breath. For example, at a depth of 30 m, the average open-circuit scuba-diver breathes at a rate of 100 l per minute, 98.9% of which is just bubbled away! In contrast, a closed-circuit rebreather (CCR) diver's metabolic consumption at 30 m is only 1 l per minute! Another way of thinking about this is to imagine a standard scuba cylinder that contains enough gas to sustain an average resting person for about 90 min at the surface. The same cylinder will last only 45 min 10 m underwater, and less than 10 min at a depth of 90 m. But if that same cylinder were filled with oxygen and used to supply a CCR, the diver could theoretically stay underwater for *2 days* – regardless of the depth!

Another advantage of CCRs is "decompression optimization". Because a CCR maintains the oxygen concentration in the breathing gas at its maximum safe value throughout the dive, the non-oxygen portion of the breathing gas (the part that determines decompression requirements) is maintained at a *minimum*. Not only does

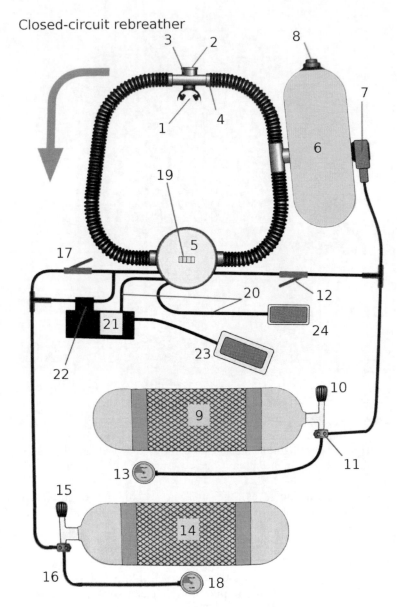

Figure 2.7. Schematic of a closed-circuit rebreather. Key: (1) Mouthpiece (2) Closing of mouthpiece (3) Return valve (to outlet) (4) Return valve (to inlet) (5) Scrubber (6) Counterlung (7) Diluent valve (8) Overpressure valve (9) Diluent cylinder (10) Diluent tap (11) Diluent control (12) Manual diluent inflator (13) Diluent manometer (14) Oxygen cylinder (15) Oxygen tap (16) Oxygen control (17) Manual oxygen inflator (18) Oxygen manometer (19) Oxygen cells (20) Cable (21) Electronic regulator (22) Electronically controlled valve (23) Primary display (24) Secondary display. Courtesy Wikimedia.

this permit the diver to stay longer at depth without incurring a decompression penalty, but it also speeds up the decompression process if a penalty is incurred.

Another drawback of open-circuit scuba is that the deeper you dive, the more rapidly you use up air, so a dive's maximum duration is determined by depth and the number of cylinders on your back. At the end of the day, a lot of the oxygen divers take with them is wasted, but rebreathers change this equation by *re-circulating* the exhaled gas for reuse and simply add a little oxygen to replace the oxygen that was consumed. The carbon dioxide is removed by a process called "scrubbing", which is achieved by an assembly that uses a soda–lime mixture (sodium hydroxide and calcium hydroxide) to absorb the carbon dioxide.

At first glance, a rebreather looks like something out of a science-fiction movie, but a closer look reveals that the system (Figure 2.7) makes sense and when you begin to understand how this system works, you being to appreciate how this different way of diving has the potential to open up new frontiers underwater. To get a better understanding, let's take a look at the components common to rebreathers.

One key element is some means to remove expired carbon dioxide from the breathing gas as it is recycled. Carbon dioxide is usually given off at a level of about 0.8 times the amount of oxygen consumed, so a rebreather has to remove about 1 l of carbon dioxide for each liter of oxygen utilized. Most rebreathers remove carbon dioxide by passing the expired gas through a canister (the scrubber, Panel 2.1) filled with chemical absorbent.

Panel 2.1. Sofnolime

Rebreathers purify breathing gas by "scrubbing". The chemical most often used to scrub the gas is soda–lime. Soda–lime – *Sofnolime* is a commercial diving product – is mostly slaked lime mixed with small amounts of more strongly based pH chemicals to help speed up the process. Some formulations also contain specific amounts of water to kick-start the chemical process. Carbon dioxide exhaled by the diver mixes with the water to form weak carbonic acid. The carbonic acid reacts with the soda–lime to form chalk, a stable solid compound that binds the carbon dioxide and removes it from the breathing gas.

Another important component is some sort of variable volume container to capture the diver's exhaled breathing gas. This is achieved by the "counterlung" (Figure 2.7), a sort of breathing bag for the diver to breathe in and out of. In addition to the counterlung, the rebreather hardware must include absorbent canisters, a means of regulating gas flow, a housing, gas storage, and a mouthpiece. Regulation of the rebreather's counterlung is affected by changes in depth. As depth changes, the rebreather unit must adjust to both a change in the gas volume and a change in the oxygen fraction in order to maintain counterlung volume and a constant partial pressure of oxygen. Consequently, ascents cause a release of bubbles and descents require addition of gas to maintain system volume, which means too many depth changes may deplete the gas supply, even if the diver does not use gas.

If you take a look at Figure 2.7, you will see the elements through which air passes during what is called the breathing loop. These elements include the mouthpiece, breathing hoses, counterlung, and scrubber (because the gas is contained inside the diver's lungs during the recycling process, the diver is also included in the loop).

While all rebreathers incorporate these key elements, there is much variability in design, each category having advantages and disadvantages, one of which is cost. A typical rebreather will set you back as much as US$20,000, not to mention the extensive additional training and certification you will need to use it. But, as we shall see, a rebreather does for scuba-diving something like a hybrid engine does for a car: it provides much greater fuel efficiency and while that may not seem like a big deal on the road, underwater it can make all the difference in the world.

Types of rebreathers

There are three main categories of rebreathers: oxygen closed-circuit, semi-closed-circuit and mixed-gas closed-circuit. Oxygen closed-circuit rebreathers, commonly referred to as oxygen rebreathers, are the most basic and least expensive rebreather design. As the designation suggests, the breathing gas is 100% oxygen and since there is no inert gas in the breathing loop, the diver does not have to worry about performing any decompression stops. In an oxygen rebreather, oxygen is added to the system through a special valve designed to maintain a constant volume of gas in the breathing loop. As the volume decreases, oxygen is added to compensate. It sounds simple, and it is, but the disadvantage of oxygen rebreathers is that the diver is breathing 100% oxygen, which means the deepest dive can be no deeper than 6 m! Deeper than this and the diver increases the risk of oxygen toxicity. However, because they are bubble-free systems, oxygen rebreathers are used extensively by the military to conduct covert shallow-water operations.

To permit divers to dive deeper, it is necessary to dilute the oxygen with an inert gas. Semi-closed rebreathers (SCRs) do this by utilizing enriched air nitrox gas (EANx) as the diluent gas. EANx is an oxygen–nitrogen blend containing a higher oxygen percentage than normal air. Although the breathing loop of an SCR is similar to an oxygen rebreather, there are some differences to accommodate the use of the EANx. One difference is the inclusion of an overpressure valve that vents gas to ensure the breathing loop is maintained at ambient pressure. Another feature is a special valve that injects the breathing gas at a concentration that overcomes the diver's metabolic oxygen consumptions. Thanks to these extra features, SCRs permit a four-fold increase in gas economy compared to open-circuit scuba and allow divers to perform dives as deep as 40 m. They are also relatively simple and inexpensive, which is why they are the most widely used rebreather by recreational divers. However, despite its versatility, the SCR design limits the diver to just 40 m, which hardly makes it a useful tool for underwater exploration. Fortunately, there is another type of rebreather that may ultimately prove to be a powerful tool in extending the range of the technical diver.

The closed-circuit rebreather

Mixed-gas CCRs are the most versatile diving units on the market today, since they offer the diver a wide depth range, long duration, optimal gas loading, and high gas efficiency. A CCR improves upon the efficiency of the SCR by injecting oxygen only when it is required to compensate for the diver's actual metabolic consumption. This is achieved by measuring the partial pressure of oxygen in the breathing loop and injecting oxygen to maintain the partial pressure. In this system, all the exhaled air is retained within a closed loop (Panel 2.2). The air is then filtered, refreshed, and recycled back to the diver for further use – a design that provides extraordinary diving endurance when compared with traditional scuba.

Panel 2.2. The closed loop

1. Exhaled gas leaves the diver's lungs and is routed into a one-way loop beginning at the unit's exhale counterlung.
2. The exhaled gas is routed, via a water-trap, into the scrubber unit, where it passes upwards through a Sofnolime filter stack, which scrubs the breathing gas of carbon dioxide.
3. Inside the mixing chamber, three independent oxygen sensors measure oxygen pressure and the partial pressure of oxygen in the gas. If the oxygen's partial pressure drops below a threshold (termed the "setpoint"), an oxygen controller opens an oxygen valve to bring the partial pressure to the threshold.
4. Scrubbed breathing gas returns via a second water-trap to the inhale counterlung ready for use in the next breathing cycle.

Because the CCR adds only a small volume of oxygen to the system at a time, the diver only has to take along a small oxygen cylinder and the only limitations on the depth and duration of a dive are the diver's metabolic rate and oxygen partial pressure preference. If the diver is diving to 46 m or shallower, the diver will take along a second cylinder holding diluent gas, while for deeper dives, they will usually use a trimix blend. Because the diver is breathing the ideal gas mixture at every depth, no-decompression limits are much longer. The only disadvantage, other than the cost (US$10,000 and up!), is the extensive training required, since divers must become intimately familiar with all the quirks and eccentricities of the particular system they are diving. The training is particularly important because the complex design means a mixed-gas CCR has the greatest risk of failure due to user error and/ or mechanical malfunction. However, recent advances in rebreather design mean that while the systems continue to be sophisticated machines, a properly trained diver can minimize the risk. A good example of such a system is the Evolution CCR.

The Evolution closed-circuit rebreather

Compact, light, and streamlined, the Evolution (Figure 2.8) is a diver's dream thanks to the system's redundancy and the future-proofed electronics package that lie at the heart of this revolutionary CCR. The Evolution's Vision versatile electronics package is as simple or as complicated as the diver wants to make it. Because it comes installed with six memorized gas mixes, it is possible to program gases for an entire diving season or simply customize them for a single dive. The computer is also programmed with Gradient Factors, enabling the diver to customize ascent profiles based on dive fitness, environment, and age. Selecting the Gradient Factors enables the diver to select a dive profile based on DCS risk. For example, if a diver selects a low Gradient Factor, by deciding to conduct deep decompression stops, he/she simply selects a number between 5 and 30 – the lower the number, the lower the risk.

Diving with a CCR as advanced as the Evolution after spending years using traditional rebreathers is akin to making the jump from flying a Cessna to flying a Citation jet. Part of the reason is the level of control that is at the diver's fingertips. In addition to a primary display and head-up display (HUD), the Evolution has built-in decompression algorithms and two independent control systems: one Master

Figure 2.8. The Evolution closed-circuit rebreather. Photo John Bantin. Courtesy Ambient Pressure Diving.

and one Slave, which serves as a backup. System redundancy is achieved by each controller having its own power supply – a feature enhanced by the operation of the Master and the Slave, since the Slave not only monitors the Master controller, but also checks the partial pressure of oxygen and gives the diver warnings if the oxygen level deviates from the setpoint. This is important because it affects the length of time a diver can actually dive. You see, the length of time a diver can spend underwater depends on a number of factors, ranging from the system's central nervous system (CNS) toxicity clock, the life of the carbon dioxide scrubber, and, of course, the gas supply. To prevent the diver from suffering the effects of oxygen toxicity, the Evolution is fitted with a default setpoint of 1.3 bar, which is limited to 3 hr per day, depending on what level the setpoint is selected. Another factor that determines the dive duration is the Sofnolime, which is depleted depending on the diver's production of carbon dioxide, water temperature, and work rate. However, even a hard-working diver using 2 l per minute would be able to dive for 5 hr using the Evolution's 2-l cylinder, which provides 400 l of oxygen.

In addition to providing extended dive duration, the Evolution also ensures a high degree of safety thanks to myriad intuitive warning systems. A solid green light indicates all systems are functioning as they should. If the diver sees any other color light, it is a warning to check the handset. In the unlikely event the handset fails, the diver can still rely on the information scrolling across the HUD, which will tell them about everything from battery life to the partial pressure of oxygen. In fact, there are three independent monitoring systems and displays available to the diver – a design that ensures a high level of system redundancy. Helping make sense of all the information is the HUD that is split into two displays (one for each oxygen controller) comprising just two pairs of lights, the brightness of which can be changed in accordance with ambient light conditions. When the system is functioning normally, the HUD shows two solid green lights, one for each oxygen controller. If the diver encounters a problem, a buzzer is accompanied by flashing red lights.

Another key safety feature is the Scrubber Monitor, which warns the diver in advance of increasing carbon dioxide concentration. The Scrubber Monitor is particularly noteworthy because until the advent of the Evolution, a reliable scrubber duration warning system had proven elusive. Utilizing an array of digital temperature sensors, the Scrubber Monitor checks the temperature profile of carbon dioxide-laden exhaled gas in real time and compares it with other sensors. This information is then displayed on the wrist-set to show the hot sections of the scrubber.

With all this information, it is hardly surprising that the Evolution (Figure 2.9 and Table 2.1) comes complete with a computer-link capability enabling divers to download up to 9 hr of recorded dive data, which include the depth–time profile, real-time partial pressure of oxygen, scrubber temperatures, surface interval, decompression, ambient temperature, and even battery voltages.

Despite their sophistication and cost, CCRs such as the Evolution have been embraced by the technical diving world because they offer a means of truly exploring the underwater frontier and, in time, they will undoubtedly become an important

Figure 2.9. The Evolution showing the harness, wings, and hose. The system is an electronics-driven fully closed-circuit constant partial pressure of oxygen rebreather that provides the diver with a host of diving information via a heads-up display. It costs about US$10,000. Photo John Bantin. Courtesy Ambient Pressure Diving.

tool for those who can afford the cost and necessary training. Yet, despite being such a powerful tool for technical divers of the future, CCRs such as the Evolution are limited to a maximum depth of 160 m.[3] While this may be deep enough even for technical divers, there are always those who really want to push the limits. For these daredevils, there is saturation diving.

SATURATION DIVING

"Saturation" simply refers to the fact that the diver's tissues have absorbed the maximum partial pressure of gas possible for that depth as a result of the diver having been exposed to breathing gas at pressure for a prolonged period. This is significant because once the tissues become saturated, the time to ascend from depth

---

[3]   On August 21st, 2003, German divers Chris Ullmann, Manfred Führmann, and Volker Clausen claimed a new depth record using closed-circuit breathing dive equipment after reportedly descending to 224.5 m.

Table 2.1. Evolution technical specifications.

Weight	26 kg	Weight of Sofnolime	2.5 kg
Absorbent duration	3 hr (4°C)	Oxygen cylinder	1 × 2 l
Diluent cylinder	1 × 2 l	Atmospheric range	650–1080 mbar
Buoyancy compensator	Wing style – 16 kg	Counterlung volume	Medium: 11.4 kg Large: 14 kg
Cylinders		Two 2-l steel cylinders: one oxygen, one diluent	
First stage (oxygen)		Intermediate pressure: 7.5–8.0 bar	
First stage (diluent)		Intermediate pressure: 9.0–9.5 bar	
Oxygen control		Two oxygen pressure setpoints	
Oxygen setpoint range (low)	0.5–0.9 bar	Oxygen setpoint range (high)	0.9–1.5 bar
Oxygen warning level (low)	0.4 bar	Oxygen warning level (high)	1.6 bar

Depth limits
Depth (m)
 40   Maximum depth with air diluent
100   Maximum depth for which all rebreather parameters proven
110   Depth at which work of breathing has been tested using trimix
150   Maximum depth at which work of breathing has been tested with heliox as diluent
160   Depth at which all components are pressure tested

and to decompress safely will not increase with further exposure. This means that SAT diving enables divers to live and work at depths greater than 50 m for days or even weeks at a time. It is a type of diving that allows for greater economy of work and enhanced safety for the divers, since, after working in the water, the divers can rest and live in a dry pressurized habitat on the ocean floor at the same pressure as the work depth. The diving team is compressed to the working pressure only once, and decompressed to surface pressure once. From an operational perspective, there are two principle factors in SAT diving: the depth at which the diver's tissues become saturated (storage depth) and the vertical range over which the diver can move (excursion depths).

SAT divers (Figure 2.10) usually live in a surface chamber known as a Deck Decompression Chamber (DDC) at a "storage" pressure that is shallower than the diver site. If the chamber is on a surface ship, the divers transfer to a Personnel Transfer Capsule (PTC) through a mating hatch in the DDC, and the PTC is lowered to the dive site. Figure 2.11 shows a typical SAT system, which usually comprises a DDC, transfer chamber, and a PTC, which is commonly referred to in

Figure 2.10. A saturation diver works outside the Aquarius habitat off the coast of Florida. Courtesy NASA.

commercial diving parlance as the diving bell. The entire system is usually placed on a ship or ocean platform and is managed from a control room, where depth, chamber atmosphere, and other system parameters are monitored.

The diving bell transfers divers from the SAT complex to the work site. Usually, it is mated to the complex utilizing a removable clamp and is separated by a trunking space, which is a kind of tunnel through which divers transfer to and from the bell. Upon completion of underwater operations, the SAT diving team is decompressed gradually back to atmospheric pressure by the slow venting of system pressure at an average of 15 m per day, traveling 24 hr a day, depending on the depth, length of time at depth, and the breathing gas. Thus, the SAT process involves only one ascent, thereby mitigating the time-consuming and comparatively risky process of in-water, staged decompression normally associated with traditional diving operations. From their SAT complex, divers use surface-supplied umbilical diving equipment,

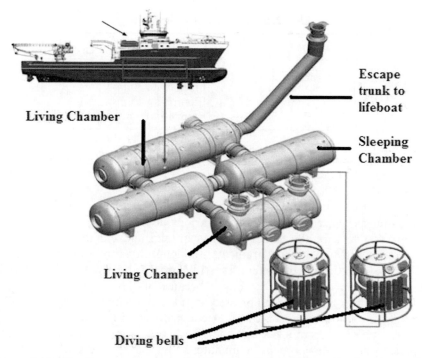

Living Chamber

Escape
trunk to
lifeboat

Sleeping
Chamber

Living Chamber

Diving bells

Figure 2.11. The layout of a typical saturation diving system showing the surface support vessel and living quarters.

utilizing breathing gases such as helium and oxygen mixtures. The gases are stored in large-capacity, high-pressure cylinders and are plumbed to the control room, where they are routed to supply the system components. The diving bell is fed via a large umbilical that supplies breathing gas, electricity, communications, and hot water. The bell is also fitted with exterior-mounted breathing gas cylinders in the event of a contingency. While in the water, the divers use a hot-water suit to protect against the cold. The hot water comes from boilers on the surface and is pumped down to the diver via the bell's umbilical and then through the diver's umbilical.

SAT diving might satisfy a diver's need for adventure but it is not a healthy way to earn a living. Long-term exposure to breathing gas under high pressure may cause aseptic bone necrosis, which is why commercial divers usually have X-rays taken at regular intervals. Then there are the dangers of breathing gas at extreme pressure to consider. For example, in the North Sea, commercial divers operate at depths of up to 500 m. At this depth, they breathe either a combination of hydrogen and oxygen (hydrox) or an exotic blend of hydrogen, helium, and oxygen (hydraliox). At these extreme depths, divers are susceptible to HPNS, insomnia, and bouts of extreme fatigue, which is the result of breathing gas that is as much as 50 times as dense as air on the surface. Divers usually describe the sensation as trying to breathe soup!

The complexity of diving to such extreme depths is compounded by the decompression requirement. For divers breathing helium–oxygen, for example,

Table 2.2. Saturation decompression.

Rate	Depths
1.8 msw/hr	From 480 to 60 msw
1.5 msw/hr	From 60 to 30 msw
1.2 msw/hr	From 30 to 15 msw
0.9 msw/hr	From 15 to 0 msw

pressure reduction must follow a very slow schedule (Table 2.2), which means for a diver working at 500 m for 2 weeks, it may take as long as a month to decompress to the surface!

Diving physiologists attempt to reduce the decompression penalty by having divers breathe specialized gas mixtures at various depths during the decompression phase, but despite carefully calculating gas uptake and elimination speeds, and despite factoring in tissue-blood gradients, the length of time to decompress from depths of several hundred meters is still measured in weeks. In fact, the decompression time may be even longer because if the diver experiences DCS symptoms, the diver must be recompressed until the symptoms are relieved and then decompression reinitiated more slowly. Decompression is thus inherently not only expensive, but also dangerous because the diver's tissues must remain continuously in a supersaturated state in order to eliminate the burden of excess diluent gas. Also, DCS cannot be predicted or prevented with absolute certainty because of its probabilistic nature and because each diver reacts differently. The same diver may have a different reaction at different times because of myriad factors ranging from dehydration to fitness and stress. Therefore, the decompression rate necessary to prevent DCS for any individual can only be an approximation based on prior general experience because all the risk factors involved in the off-gassing rate for a given person are not and can never be known. Inevitably, a method for shortening decompression would reduce a time of great personal risk to the diver as well as reducing expenses of the dive operation. Such a method would represent the future of SAT diving and it is discussed here.

Biochemical decompression

The notion of biochemical decompression was first proposed by Dr. Lutz Kiesow, a leading scientist at the Naval Medical Research Institute (NMRI) who suggested using hydrogenase to cause biochemical decompression. According to Dr. Kiesow's idea, a diver would breathe a gas blend containing hydrogen and oxygen. The diver would then be supplemented with a hydrogenase enzyme, which is found in some bacteria. The hydrogenase enzyme would convert the hydrogen to some other molecule – a process that would ameliorate the diver's burden of excess diluent gas during the ascent phase, thereby reducing the decompression time. Subsequent research attempted to put Dr. Kiesow's concept into action by putting purified hydrogenase into red blood cells directly. The scientists found it was possible to encapsulate the enzyme into red blood cells, but they could not devise an animal model in which to test the cells. The concept of incorporating the hydrogenase into the blood was not pursued because even if the enzyme could be packed into the blood cells and be injected into a diver, the cells could only be circulated for a few weeks before the red blood cells died naturally and were eliminated through the spleen. Another problem was the foreign protein of the injected enzyme could lead to splenic failure!

More recent research [1] has investigated Dr. Kiesow's concept by introducing into the large intestine a non-toxic bacteria selected from the group that metabolizes hydrogen. As part of her work for the Office of Naval Research (ONR), Susan Kayar performed studies on deep-diving hydrogen-breathing pigs that had been fed the hydrogen-metabolizing bacteria [2, 3]. During the dive, the bacteria multiplied and fed on the gas mixture used in the dive by metabolizing the diluent gas released into the large intestine. The new product was simply vented from the large intestine. The metabolism of the hydrogen caused a reduction in the partial pressure of the metabolized gas in the large intestine, thereby increasing the diffusion of the metabolized gas from the blood and surrounding tissues into the intestine. To deliver the bacteria to the pig's intestines, Kayar simply packaged it in an enteric coating for oral ingestion. The coating was necessary to protect the bacteria while passing through the stomach. By the time the package reached the large intestine, it had dissolved and released the bacteria.

Since Kayar's invention [4] has the potential to revolutionize saturation diving, it is worthwhile examining some of the principles that underlie the process. The first step was to find bacteria capable of metabolizing hydrogen into methane but, naturally, the bacteria could not be toxic to the diver. Fortunately, Kayar found several bacteria that met the requirements for bacterial decompression, one of which was *Methanobrevibacter smithii*. The next challenge was to ensure the bacteria were capable of returning to the active metabolism. One of the preferred solutions was to wrap the bacteria in a slow-release capsule as a freeze-dried product. The intention was the diver would simply swallow one or more capsules containing the bacteria. The packaging would pass through the stomach and small intestine unharmed. Once in the small intestine, the capsule would begin to dissolve and would be fully hydrated and operational by the time it reached the large intestine, allowing the bacteria to colonize there indefinitely. To provide sufficient lead time for the bacteria to reach and colonize the large intestine, the diver would be required to ingest the capsule about 12 hr before the bacteria were needed to assist in decompression.

Conceptually, Kayar's invention offered the possibility of dramatically reducing decompression time. But it had not been tested. One of the first studies placed hydrogen-metabolizing bacteria inside the large intestines of rats (Panel 2.3). Sure enough, the hydrogen-metabolizing bacteria did indeed eliminate hydrogen dissolved in the rat's tissues and the rat's risk of DCS was subsequently lower.

How did the bacteria do this? The culture of bacteria that was introduced into the large intestines of the lab rats possessed an enzyme of the class known as hydrogenase, which is a protein enzyme that catalyzes the metabolism of hydrogen.

Panel 2.3. Hydrogen-breathing lab rats

To test the viability of hydrogen-metabolizing bacteria as a means of reducing decompression, live cultures of *Methanobrevibacter smithii* in a bicarbonate buffer were surgically injected into the proximal end of the large intestines of five rats via a cannula. The rats were placed in a box inside a dive chamber that was specially designed for operating with gas mixtures of hydrogen and oxygen and exposed to 100 m of sea water equivalent pressure of a hydrogen–oxygen mixture.

A stream of gas passing through the box was sampled by gas chromatography. The animals were then placed in a dry hyperbaric chamber specially designed for compression with mixtures of hydrogen and oxygen. As the rats breathed hydrogen, some of the gas was metabolized by the microbes in the intestines while the chamber gases were monitored by gas chromatography. The rate at which the rats released methane was measured by following the rate of hydrogen removal by the microbes. The study demonstrated that the rats that received the microbial treatments had a significantly lower incidence of DCS compared to untreated animals, and also compared to surgical control rats that received intestinal injections of saline. In fact, the *Methanobrevibacter smithii* cut the incidence of DCS by half.

The category of hydrogenase chosen for the biochemical decompression study was the methanogen, a hydrogenase that metabolizes hydrogen to form methane. Of the methanogens studied, *Methanobrevibacter smithii* was an ideal candidate for the purposes of biochemical decompression because this species is a common resident of the normal human gut flora and has no known pathogenicity. During its action, *Methanobrevibacter smithii* converts hydrogen and carbon dioxide to methane and water, consuming four hydrogen molecules for each molecule of methane produced. Under normal circumstances, the source of hydrogen for this reaction is the end-product of fermentation by other bacteria in the intestine. While most people on a Western diet usually harbor only small populations of *Methanobrevibacter smithii* and produce milliliter volumes of methane per day, there are some healthy individuals who produce up to 4 l of methane per day, thus metabolizing 16 l of hydrogen! The methane passes harmlessly from the rectum.

Calculating the amount of bacteria can be estimated by factoring in the length of activity, the volume of hydrogen to be consumed, and the length of time by which the decompression time is to be reduced. To determine how much bacteria the diver must ingest requires certain assumptions. First, it is assumed that the partial pressure of gas in the blood stream is supersaturated as the diver decompresses and no additional gas will dissolve into the blood stream. Second, it is known that the total volume of hydrogen in a diver per unit body mass is a linear function of the partial pressure of hydrogen to which the diver is exposed. Therefore, if a diver is at maximal depth and in a steady state, saturated with hydrogen, and sufficient bacteria were present to consume 50% of the body burden of hydrogen over the same time interval at which 50% of their body burden of hydrogen would normally be offloaded by traditional decompression procedures, the overall speed of decompression would be doubled, and the time to decompress would be halved. Of course, if a diver wanted to cut even more off his/her decompression time, he/she could be supplemented with greater quantities of bacteria, to remove even more hydrogen per unit time. Popping a few capsules containing bacteria certainly beats hanging onto a decompression line breathing mixed gas for hours at a time!

Kayar's research represents a fundamentally different approach to reducing DCS risk. Whereas previous research has sought to adjust dive duration and depth combinations, Kayar's work seeks to reduce DCS risk by actively eliminating a critical portion of the body's inert gas load. It is a radical approach, but research to date indicates that if divers are provided with the biochemical machinery to metabolize hydrogen saturated in their tissues, a significant reduction in DCS may be achieved.

Only a few years ago, a discussion of non-commercial divers using mixed-gas equipment would have been more than a little far-fetched to many diving professionals, and to suggest that safely engineered CCRs could be available off the shelf would have caused many to choke on their regulators. Similarly, if you had explained to the average technical diver a couple of decades ago that it would be possible to pop a pill that would halve their decompression time, you would probably have received some strange looks. Nevertheless, here we are in 2010, with technological innovations that herald the future of technical and SAT diving. However, while CCRs probably represent the future of scuba and while biochemical decompression may well reduce the risk of DCS, even with these advances, humans will still be restricted to depths in the 100–200-m range and they will still be required to undergo at least some decompression. The average depth of the ocean, however, is 3,790 m, which means using the most advanced SAT systems and most cutting-edge diving equipment, humans will still be limited to barely scratching the surface of the planet's least explored region. To embark upon real underwater adventure requires a more robust diving system and it is this that is the subject of the next chapter.

REFERENCES

[1]   Fahlman, A.; Tikuisis, P.; Himm, J.F.; Weathersby, P.K.; Kayar, S.R. On the Likelihood of Decompression Sickness During H (2) Biochemical Decompression in Pigs. *Journal of Applied Physiology*, **91**(6), 2720–2729 (2001).

[2]   Kayar, S.R.; Fahlman, A. Decompression Sickness Risk Reduced by Native Intestinal Flora in Pigs after H2 Dives. Naval Medical Research Center, Silver Spring, Maryland, USA. *Undersea Hyperb Med.* **28**(2), 55–56 (2001).

[3]   Kayar, S.R.; Miller, T.L.; Wolin, M.J.; Aukhert, E.O.; Axley, M.J.; Kiesow, L.A. Decompression Sickness Risk in Rats by Microbial Removal of Dissolved

Gas. *American Journal of Physiology – Regulatory, Integrative and Comparative Physiology*, **275**, R677–R682 (1998).

[4] Kayar, S.R.; Axley, M.J. (Inventors). Accelerated Gas Removal From Divers' Tissues Utilizing Gas Metabolizing Bacteria. US Patent 5,630,410, May 20, 1997.

Section II

Manned Submersibles and Undersea Habitats

By generous estimates, man has explored only 5% of the ocean, and most of that has been within 300 m of the surface. This shallow part of the sea is the familiar blue portion, penetrated by sunlight, home to colorful reefs and most of the world's fish. Since the lower limit of most of the world's deep-diving submersibles extends to only 6,000 m, it isn't surprising that most of the major oceanographic discoveries have been made in this region of the sea. But, in their quest to venture ever deeper, engineers are designing a new fleet of manned submersibles that will eventually take humans to the Challenger Deep, the deepest part of the ocean – a place not visited since January 23rd, 1960.

Here, in Section II, we delve into the new technology that is changing the way engineers design submersibles. First, in Chapter 3, we discuss the future of the one-man submersible, commonly known as the hardsuit. Then, in Chapter 4, we take a look at the work of the science submersibles such as Alvin, a vehicle that has enabled true exploration of the world's deepest waters. In Chapter 5, we discover how Graham Hawkes is revolutionizing underwater exploration by designing aircraft-like submersibles capable of executing barrel rolls and diving to depths not reached in half a century. Finally, in Chapter 6, a vision of a permanent underwater base is discussed and the reader is introduced to the world of the aquanaut and the challenges of living underwater.

3

Hardsuits

While diving to depths between 200 and 500 m can be achieved by freediving, technical diving or saturation (SAT) diving, for deeper submersible operations, a different system is required. An atmospheric diving suit (ADS) is an articulated one-man, anthropomorphic, articulated submersible (Figure 3.1) that bears more than a passing resemblance to a suit of armor. Elaborate pressure joints permit articulation while maintaining an internal pressure of 1 atmosphere. Thanks to its design, the ADS can be used for very deep dives of up to 1,000 m or more for many hours, and eliminates most physiological dangers associated with technical and SAT diving such as decompression sickness, nitrogen narcosis, and breathing exotic mixed-gas mixtures. As we saw in Chapter 2, SAT dives change the way gases dissolve in the divers' blood and high pressures below the surface force breathing gases into solution with the blood and body tissues, requiring SAT divers to surface slowly. As they ascend and the pressure around them drops, dissolved gas comes out of solution and if the ascent is slow enough, reformed gas bubbles leave through the lungs. But ascending too fast traps these bubbles in body tissues, causing joint pain, neurological effects, shortness of breath, and even death. For example, a conventional SAT diver would need to undergo a slow, finely controlled decompression lasting between 8 and 10 days after he/she had acclimated to pressures equivalent to 300 m below the surface. By comparison, atmospheric dives, like those performed in modern ADS systems, require only 3–5 min recovery, independently of dive depth and duration, which means the ADS suit can venture up and down the water column any number of times without consequence or delay.

Another drawback with SAT diving is cost. In addition to a crew of 12 or more, SAT living chambers and the diving bell that transports saturated divers between living chambers and the worksite eat up several hundred square meters of critical deck space. With gas storage, a typical SAT system can weigh upwards of 100 tonnes and cost US$5 million or more. Even the cost of the gas may run into the hundreds of thousands of dollars. In contrast, operating ADS systems requires only a four-person crew and the suit itself weighs less than 1 tonne and takes up only a few square meters of deck space. Diving gas costs about US$50 a dive, and the suit itself costs less than a million dollars. Hardly surprising, then, that organizations around

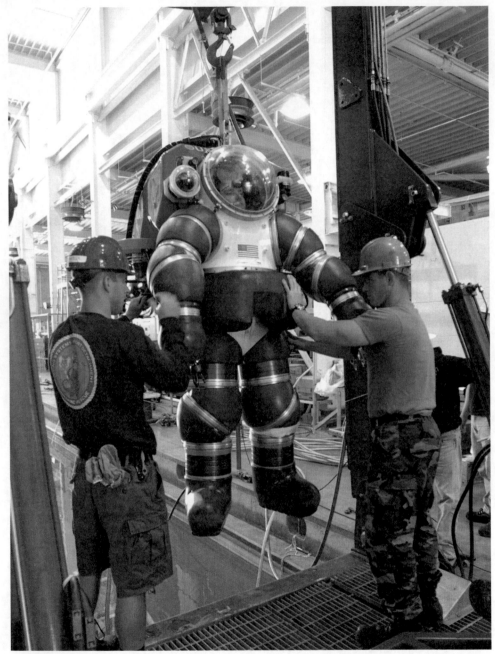

Figure 3.1. An atmospheric diving suit (ADS) is a one-man articulated submersible with articulated pressure joints. Some ADS systems can be used for dives of up to 700 m, and because the suit is maintained at 1 atmosphere, the dangers of decompression sickness are eliminated. ADS systems currently in use include the Newtsuit, the WASP, and the ADS2000. Courtesy United States Navy.

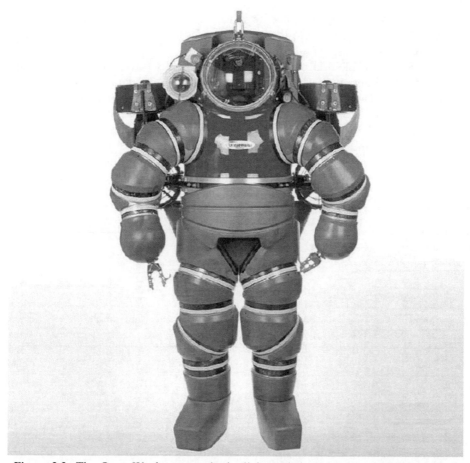

Figure 3.2. The OceanWorks atmospheric diving suit includes advanced life support monitoring and communications systems, which ensure diver safety. The life support system is a self-contained rebreather system that uses a carbon dioxide scrubber in conjunction with an onboard oxygen supply. The suit has 16 rotary joints and two foot-pedal-controlled thrusters to provide mobility. Its primary mission is submarine rescue. Courtesy OceanWorks.

the world, including the military and oil and gas exploration companies, use ADSs for tasks ranging from aircraft and weapons recovery to search and salvage and from submarine escape to inspection and maintenance of sub-sea platforms.

ADS2000 HARDSUIT

On August 12th, 2000, the Russian Oscar II class submarine *Kursk* sank in the Barents Sea and despite a rescue attempt by British and Norwegian teams, all 118 sailors and officers aboard died. The *Kursk* tragedy resulted in a dramatic increase in interest in ADS technology, an outcome of which was the implementation of the US Navy ADS2000 (Figure 3.2) into the Submarine Rescue Diving and Recompression System (SRDRS).

The ADS2000 (Table 3.1) was built by OceanWorks International Corporation (OWC) of North Vancouver, Canada. The fact a Canadian company was selected by the US Navy to design and build a system intended for use as part of a submarine rescue system was ironic, since the Canadian Navy has no submarine rescue system of their own!

Using cast aluminum pressure hull components, OWC was able to produce the complex curved shapes necessary to accommodate the ergonomic requirements

Table 3.1. ADS2000 technical summary.

Overview	
Maximum depth	610 m (working pressure: 890 psig)[1]
Height	2.42 m
Width	1.22 m
Weight	518 kg (without pilot)
Hull material	6061 T651 forged aluminum
Vision dome	Reynolds Polymer Technology R-Cast acrylic window
Fixed buoyancy	Floatec syntactic foam
Propulsion system	
Power	4 × 1.1-hp thrusters
Control	Foot controls
	Left foot: vertical control
	Right foot: lateral control
Communication and navigation equipment	
Hard wire	Digital voiceover data – Seaphone Model 3700
Through water	25 or 8 kHz acoustic – underwater telephone transducer
Emergency communication	Radio frequency beacon/Xenon strobe light with pressure switch Emergency 37.5 kHz pinger
Navigation and vision systems	Navigation: color imaging sonar, digital compass, altimeter and acoustic positioning system
	Vision: external lights (two) Deep Sea Power and Light Mini Sealight
Life support	
System	Oxygen recirculation with fan-powered carbon dioxide scrubbing capability
	Dual independent oxygen system – port and starboard
Duration	6–8 hr with 48-hr emergency life support

[1] Psig (pound-force per square inch gauge) is a unit of pressure relative to the surrounding atmosphere. By contrast, psia (pound-force per square inch absolute) measures pressure relative to a vacuum (such as that in space).

required by operators working with equipment several hundred meters below the surface.

The ADS2000 thruster design uses commercial off-the-shelf (COTS) DC brushless thrusters combined with a microprocessor control system and a top-side power delivery system that transmits high-voltage DC via the umbilical and tether. The microprocessor control system provides the operator and the supervisor with control of all thruster settings, including maximum thrust and trim. The microprocessor also provides troubleshooting and diagnostics information to the supervisor.

Designed to provide rapid mobilization for any operational sub-sea task, including disabled submarines (DISSUBs, Panel 3.1) and salvage operations, the ADS2000 can be easily airlifted to any location and installed on a vessel of opportunity (VOI). At first glance, you might think the operator would be encumbered by the bulkiness of the design, but you would be wrong. The suit's

---

Panel 3.1. Submarine escape

There are two approaches to extracting crewmembers from a DISSUB. The first is for personnel themselves to make their way to the surface using breathing gear and the vessel's escape hatches. The second option is for a rescue vessel to mate with the downed sub and transfer the stranded personnel to the surface. The rescue vessel may be either a manned or an unmanned submersible. For example, the US has two Deep Submergence Rescue Vehicles (DSRVs), one stationed in the Atlantic and one in the Pacific. The DSRVs have a crew of four and can hold 24 rescued crewmembers. There are also Submarine Rescue Chambers (SRCs), which are pressurized diving bells designed to be lowered onto a submarine from a ship. An SRC can transport six evacuees to the surface.

Both systems utilize a skirt that covers the hatch of the DISSUB. When sealed to the hull, the skirt acts as an airlock, pumping out water and equalizing pressure between the two vessels before the hatches are opened. Both systems require support from divers and/or remotely operated vehicles (ROVs). In the case of the SRCs, cables must be tethered to the damaged sub so the chamber can stay on target.

With the ADS2000, there is now a third option for rescuing stricken submariners. Here is how it would work: once a DISSUB has been located, an ADS2000 team would be deployed to the site and begin conducting an initial survey, providing rescuers with video, sonar, and onsite reports. The primary task of the ADS2000 operators would be to clear debris from the submarine hatch, remove the hatch fairing, and connect the downhaul cable for the SRC. Thanks to the suit's life support system, the ADS2000 team could work at depths of 600 m for up to 6 hr, with additional emergency life support for up to 48 hr.

rotary joints allow the operator to perform movements using a full range of motion, and the thruster pack allows work in mid-water, enabling the operator to work in just about any position, from fully upright to fully prone. The range and maneuverability of the suit also enable access to restricted spaces to perform work not accessible by submersibles and previously the exclusive domain of SAT divers.

The ADS2000's life support system (LSS) is a self-contained rebreather system that uses a carbon dioxide scrubber in conjunction with an onboard oxygen supply. The oxygen supply is fully redundant, with two pairs of oxygen bottles, two regulators, and two internal bellows-activated control systems. During normal operations, one system is in use and the second system is held in reserve. An important safety feature of the ADS2000 is a cutting-edge Atmospheric Monitoring System (AMS), which provides the dive supervisor with continuous monitoring of all ADS2000's internal conditions, including oxygen percentage, carbon dioxide concentration, cabin pressure, depth, as well as cabin and water temperature. In addition to atmospheric monitoring, the AMS surface display also provides the supervisor with feedback about all suit systems, including microprocessor and power supply status, troubleshooting aids, and compass heading, thereby reducing the risk of an incident during a dive.

During dive operations, the ADS2000 operator is in contact with a support team located in a control room, which is divided into two areas: a workshop and a control room. The control room contains a console, which provides mounting for the ADS2000 surface equipment and the Launch and Recovery System (LARS) control module. In August 2006, the ADS2000 system was put through its paces when Chief Navy Diver, Daniel P. Jackson (Figure 3.3), submerged 600 m off the coast of La Jolla, California:

> "The suit worked incredibly. It did everything it was intended to do. I always heard that around 1,300 feet, the joints of the Hardsuit 2000 would work even better, and it worked exactly the way they said it would."
>
> Chief Navy Diver, Daniel P. Jackson,
> after putting the Hardsuit 2000 through its paces

Training

In common with most advanced diving activities, becoming certified in the ADS2000's operation requires training and a thick wallet. Enrolling in the 5-week course will cost you US$22,000, although this fee does include the hire of a pilot's kit, personal protective equipment and clothing, and operators' manuals.

The training begins with the inevitable Powerpoint presentations, which detail aspects relevant to the operational capabilities of the ADS2000 and provides an overview of the resources available to make the most of the suit. During the practical sessions, the pilot trainee is expected to perform a number of tasks that reflect situations and operations that may be encountered in the field. Most pilots require 20 hr to learn basic skills and become proficient with the suit's life support and flying

Figure 3.3. Chief Navy Diver Daniel P. Jackson is hoisted onboard the *M/V Kellie Chouest* following a successful certification dive of the Atmospheric Diving System (ADS 2000) off the coast of La Jolla, California, in August, 2006. Jackson's 600-m dive set a US Navy diving record. US Navy photo by Mass Communication Specialist Seaman Chelsea Kennedy. Courtesy United States Navy.

functions to allow them to focus on the work tasks. During the course, pilot trainees are also introduced to the various aspects of job preparation, planning, and tool modification required for use with the ADS2000. For example, typical tasks required to be completed by students include sizing up and buoyancy set-up, overall operating procedures in-water exercises, emergency procedures in-water exercises, and "flying" in and around structures. They also have to learn how to monitor life support system readings while hovering and maintain through-water communications. Control of the suit is achieved by pedals in the feet of the suit that control four back-mounted thrusters. By pressing up or down on the right foot, the operator controls the horizontal motion of the suit and by pressing up or down on the left foot pedal, they control the vertical motion of the suit. By pressing on the instep or outstep of the right pedal, it is possible to spin around the vertical axis of the suit. It may sound difficult, but at the end of the 20 or so hours of training inside the suit, most operators are dexterous enough to pick a quarter from the bottom of the pool!

While the ADS2000 is undoubtedly an advanced and versatile suit, the operator must still rely on surface support, which is an encumbrance when dealing with complex tasks such as submarine rescue. To fully realize the potential of the hardsuit, a system should be swimmable – a requirement that leads us to a system that may represent the future of ADS systems: the Exosuit.

EXOSUIT

Firmly at the cutting edge of ADS technology is the Exosuit (Figure 3.4). Envisioned by legendary underwater inventor and pioneer, Dr. Phil Nuytten (Panel 3.2), and built by Nuytco Research Ltd of North Vancouver, Canada, the swimmable Exosuit is so agile that operators describe its operation as akin to wearing a submarine.

In common with all ADS systems, the Exosuit's rigid structure withstands immense water pressure thanks to its A356-T6 aluminum alloy cast skin. Rated to 365 m and an impressive crush depth of more than 700 m, the Exosuit is able to support operations lasting up to 3 days. The core of the suit encases the diver's torso and head while rotary jointed arms and legs permit easy movement. Although the suit's arms and legs are heavy in air — the suit weighs more than 100 kg on land – in the water, the force required to move it is similar to that required by a spacewalking astronaut working in a spacesuit. The ease of movement is thanks to the confined volume of air combined with a foam flotation coating, which means the limbs remain neutrally buoyant in water. The enhanced limb mobility enables divers to propel themselves by walking or swimming using the suit's swim-fin boots, although a thruster pack that will allow the diver to "fly" through the water is being developed.

The arms are perhaps the most versatile part of the suit. Instead of utilizing gloves, like spacesuits, the Exosuit uses an artificial hand called the Prehensor, which has three fingers that mimic human hand movements, permitting Exosuit divers to retain up to 95% of their manual dexterity. This is a big improvement on previous ADS systems that traditionally used manipulators with two jaws that worked like pliers, which divers operated by squeezing handles inside the suit. It is a capability

Figure 3.4. Exosuit is a swimmable atmospheric diving suit that enables divers to dive to 365 m while breathing air at atmospheric pressure for up to 3 days. A saturation diver diving to such depths would need to spend 8–10 days decompressing afterwards, whereas divers using the Exosuit require only 3–5 min recovery, independently of dive depth and duration. Covered in an Λ356-T6 aluminium alloy skin, the Exosuit is rated to 365 m, although its crush depth is over 600 m. Weighing 118 kg, the suit appears cumbersome, but appearances can be deceiving; the suit is extraordinarily mobile thanks to special joints and Prehensors that almost exactly mimic hand movements. Whereas a saturation chamber, bell, and gas storage might cost in excess of $4 million, the Exosuit costs just $250,000. One version of the suit is designed for submarine rescue. Surprisingly, despite being Canadian-built by Nuyttco in North Vancouver, the Canadian Navy has no submarine rescue capability and has expressed no interested in acquiring the Exosuit. Courtesy Nuyttco.

Panel 3.2. Phil Nuytten, OBC, L.L.D., DSc.

Dr. Phil Nuytten has spent his life devoted to ocean exploration and is widely regarded as one of the pioneers of the modern commercial diving industry and a significant force in the realization and implementation of new diving technology.

In the 1960s and 1970s, Dr. Nuytten was involved in experimental deep diving and the development of mixed-gas decompression tables – research that helped establish many of the international diving standards in use today. During this period, Dr. Nuytten also co-founded Oceaneering International Inc., one of the largest underwater skills companies in the world.

One of Dr. Nuytten's most important contributions to ocean exploration has been the development of ADS technology. In 1979, he began developing a revolutionary ADS that resulted in a patented breakthrough in rotary joint design, and formed the basis for the world famous Newtsuit that is now standard equipment in many of the world's navies, except the Canadian. In 2000, Dr. Nuytten introduced the Exosuit, which was investigated by the Canadian Department of National Defence (DND) as a submarine escape device. Plans to utilize a space version of the Exosuit are also under discussion.

---

that has attracted the attention of NASA, as well as the Canadian Space Agency (CSA), as a possible alternative to conventional spacesuit gloves.

Exosuit operators breathe a standard 80% nitrogen, 20% oxygen mix at atmospheric pressure. The gas is supplied to match the exact amount metabolized by the operator and a soda–lime chemical absorbent scrubber removes exhaled carbon dioxide from inside the suit. Although the scrubber and oxygen circulator are battery-powered, carbon dioxide can also be removed by passive scrubbing if the battery fails. While operating the suit, the diver is tethered to the surface and lines supply him with breathing gas. Although the Exosuit is currently a prototype bare-bones version, one variant may be designed to allow submarine crews to escape disabled vessels. When a submarine loses power, it also loses the ability to keep its internal atmosphere at surface conditions, which means the crew's bodies become saturated, just like a SAT diver. If this happens, crewmembers can don suits to keep their tissues from getting saturated with atmospheric gases while they repair the sub or wait for rescue. An alternative is to allow their bodies to gas-saturate. Then, as they make their ascent, the Exosuit could maintain the pressure they had experienced on the sea floor and once on the surface, the Exosuit could serve as a personal decompression chamber for each crewmember. Thanks to its 22 highly mobile rotary joints, the Exosuit would provide crewmembers with enough flexibility to climb up into a stricken sub's escape lock, operate valves, and perform other tasks required to perform an emergency ascent to the surface.

THE FUTURE OF ATMOSPHERIC DIVING SYSTEMS

In addition to serving as a future submarine rescue system for navies around the world, the ADS is still evolving to meet the demands of the oil industry, which is poised to go ever deeper in search of oil. However, with the availability of alternative underwater intervention methods, especially those that can go deeper and without the risk to human life, no matter how small that risk may be, there are many who argue that there is no requirement for ADS systems to go deeper. However, given the current demands of the offshore oil and gas industry, the demand for ADS systems will continue to grow for the foreseeable future. These demands will require ADS systems to extend their capabilities, albeit within current depth limitations, by adding work packages that will allow them to perform a greater variety of tasks. For example, the requirements for inspection of fixed platforms in the outer continental shelf (OCS) area are determined by the Minerals Management Service (MMS), which states that "All platforms installed in the OCS shall be inspected periodically". At last count, the MMS website reported 3,412 active platforms in the Gulf of Mexico installed at depths of 200 m or less. Increased environmental concerns coupled with the increasing age of these shallow-water platforms and mounting concerns over storm damage have led to significant growth in the market for inspection services. As demand increases, oil companies are searching for inspection techniques that provide accurate and cost-effective inspections. The traditional approach to platform inspection has been to use either remotely operated vehicles (ROVs) or divers. An ROV equipped with cameras, water-jetting, and cleaning tools can be used for platform inspection, since it has the advantage of high speed in the water column, practically unlimited depth capability and the ability to make long horizontal excursions (Table 3.2). However, the shortcomings of the ROV include poor access to structures, limited visual acuity, and limited spatial awareness. Another disadvantage of the ROV is that it requires highly adapted non-destructive testing (NDT) tools and equipment for remote operation. In contrast, the modern ADSs combine the advantages of ROVs and divers with few, if any, of the disadvantages. The Newtsuit and ADS2000 are capable of using hand-held diver NDT tools with few, if any, modifications. As we have seen, by using modern ADS systems, decompression considerations are eliminated and horizontal/vertical excursion capability is comparable to that achieved by an ROV. Furthermore, access to structures and visual acuity (Table 3.2) is comparable to divers, although equipment and operating costs are significantly lower than saturation diving. Finally, the cost of placing an ADS does not increase greatly with depth. Basically, it does not cost the operator much more to send the ADS down a few extra hundred meters, whereas SAT diving costs are directly related to depth.

Just as critics condemned putting the first man on the Moon, skeptics of hardsuits ask: "If we can do it with robots, why risk a human life?" It is not a simple question to answer. ROVs have certainly surpassed hardsuit systems in their ability to go deeper and, despite a nearly flawless safety record, the risk inherent with the hardsuit *does* involve human life – something that is clearly not a factor during ROV operations. Of course, there is a risk any time a diver enters the water because

Table 3.2. Capabilities comparison for platform inspection.

	ROV	Air diving	Saturation diving	Hardsuit TM
Range				
Depth capability	> 1,000 m	< 60 m	300 m	400 m
Horizontal excursion	Excellent	Moderate	Poor	Excellent
Vertical excursion	Fast	Moderate	Slow	Fast
Economics				
Capital equipment cost	Moderate	Small	Low	Moderate
Crew cost	High	High	Moderate	Moderate
Capability				
Still photography	Yes	Yes	Yes	Yes
Visual acuity	Moderate	Excellent	Excellent	Excellent
Marine growth removal	Slow	Fast	Fast	Fast

humans were simply not designed to be subjected to extreme pressures. But, as has been proven, with the proper redundancy and safety features, the hardsuit can provide a safe work environment for the human operator. While ROVs can undoubtedly perform many of the same tasks deeper than a hardsuit, the unmanned option is not always faster or more cost-effective because machines simply cannot cope with unexpected situations or rapidly changing environments. Just ask a hardsuit operator who goes to the rescue when an ROV becomes entangled on the seabed!

So, where do hardsuits go from here? Well, almost certainly, they will be involved in continuing to extend the human diving envelope thanks to the demand for oil. Over the past few years, the oil companies have steadily pushed to develop reserves in deeper waters. According to the 2000–2004 World Deepwater Report, prior to 1960, 60 m was the maximum depth from which oil and gas were produced. By 1990, this depth had passed 600 m. This trend, coupled with the high productivity of many deepwater fields, is generating a frenzy for deeper developments. It was a similar trend towards deepwater in the 1960s that renewed the lagging interest in hardsuits. Unlike the 1960s, the oil and gas industry have alternatives such as the ROV, but this does not mean there won't be developments in hardsuit technology. This is because the availability of current hardsuits such as the ADS2000 and future systems are likely to influence engineering decision making. Already, the ADS2000 has been tested to 900 m at Carderock Naval Surface Warfare Center and the suit's designers are confident the system can be operated as deep as 2,250 m. While the future of the hardsuit seems assured in the oil and gas industry, the technology seems equally accepted by the navies of the world, with the French, US, and Italian navies all owning hardsuits as part of their submarine rescue programs. But, while the hardsuit's future seems all but assured, there is one question that remains to be answered: how will current hardsuits change to increase their performance envelope?

First, hardsuit manufacturers will almost certainly improve the battery performance, enabling them to work longer at depth. Also, as seen with the design of the Exosuit, there will be a move towards making manipulators more hand-like and providing operators with tactile feedback by means of pressure sensors. In common with technical diving systems, a heads-up display (HUD) will come as standard in future hardsuits. These HUDs will display not only oxygen and depth information, but also information generated by special shape-sensing sonar systems enabling the operator to see in the dark. In terms of depth, hardsuit performance will be dictated by the requirements of the oil and gas industry; if there is an oil field at 3,000 m, the chances are a hardsuit will be developed to work down there!

4

Manned Submersibles

"Would you send a robot to taste wine in Paris?"
Celebrated oceanographer, Sylvia Earle, when asked whether deep-sea
exploration should be conducted using remotely operated
vehicles or manned submersibles

The vehicle that enables true exploration of the deepest waters of the ocean is the manned submersible. The submersible is akin to a small submarine, but submarines are not submersibles. Submarines are designed for human occupancy and machines, usually with a military purpose, that can survive at depth for an extended length of time. For example, some nuclear-powered submarines stay submerged for months, carry food and fresh water for crews of over 100 persons, purify air for breathing, and perform warfare, espionage, and research tasks. While traditional submarines also have highly sophisticated equipment, including sounding devices, and elaborate navigation and power systems, these are used for different purposes than the instruments on a submersible, which typically include mechanical manipulators, television systems, cameras, and an array of special lighting systems. Thanks to the number of systems onboard, the manned submersible is a much more versatile craft than a submarine, as reflected by the myriad applications of the vehicle (Table 4.1). Submersibles are also designed to dive to much greater depths than submarines. Because of the tremendous pressures in the deep ocean, they are built for strength, and usually designed to carry no more than two or three human occupants, limited stores of food and water, and oxygen furnished from limited onboard storage tanks.

MIR

The Mir submersibles (Figure 4.1 and Table 4.2) were designed and built by scientists at the Russian Academy of Sciences and the Finnish company Rauma Raepola Oy. In common with their American counterpart, Alvin, the two Mir submersibles, *Mir 1* and *Mir 2*, represent the cutting edge of underwater exploration, capable of diving to a depth of almost 6,000 m (equivalent to two-thirds the height of Everest), allowing them to explore 98% of the ocean floor.

Table 4.1. Applications of manned submersibles.

Naval and corporate offshore applications
- Inspections of pipelines and communications cables
- Precision sea floor mapping for cable and hydrographic surveys
- Logistical support for cable, pipeline, and submerged structure installations
- Survey of well and dumping sites for environmentally critical wastes
- Preparation and performance of salvage operations
- Covert harbor, coastline, and offshore structure monitoring and security
- Documentary and motion picture filming
- New system testing before the installation on military vessels
- Naval sonar practices and ASW exercises

Research and educational applications
- Marine biology
- Marine archeology
- Eco-system monitoring
- Sub-sea technologies
- Marine habitat studies
- Migration studies

Table 4.2. Mir technical summary.

Length	7.8 m
Height	3.0 m
Weight	18,600 kg (maximum payload: 290 kg)
Power	100-kWh NiCad batteries
Communication	VHF radio
Life support	246 man-hour capacity (3.42 days for three-person crew)
Depth	6,000 m (field-tested to 6,170 m)
Vertical speed	40 m/min

Designed for scientific research, the Mir can accommodate a crew of three for up to 3 days, although it is unlikely the crew would want to test this capability, since there are no toilet facilities onboard! Weighing a little over 18 tonnes, the Mir's 2.1-m-diameter cabin sphere is constructed of specially strengthened nickel steel and, at 5 cm thick, is designed to withstand the enormous pressures encountered several kilometers below the surface. Diving the submersible to these depths is a slow process, as the descent rate is typically only 35–40 m/min, which means reaching a depth of 6,000 m can take as long as 2 hr! Once on the seabed, the pace hardly changes, as the Mir travels around at a top speed of 5 knots, propelled by propellers mounted on the side of the submersible. As the submersible explores the seabed, scientists can observe through multiple view ports, enabling them to place instruments, collect samples, and monitor the environment. However, scientists are not the only ones to employ the Mirs; Hollywood director James Cameron used them to make his blockbuster *Titanic*, and they have also been used for IMAX films.

Figure 4.1. The Mir submersible is lowered into the water by a cable connected to the ship's winch. This view shows the versatile manipulator arms and the huge viewing port. The Mir project was developed by the USSR Academy of Sciences along with Design Bureau Lazurith. Built in 1987, *Mir 1* and *Mir 2* were designed and built by the Finnish company Rauma, Raepola's Oceanics subsidiary. In the mid 1990s and early 2000s, the submersibles were used by Canadian film director, James Cameron, to film the wreck of *Titanic*. On August 2nd, 2007, Russia used the submersibles to descend to the seabed under the Geographic North Pole (a depth of 4,261 m), a stunt that featured *Mir 1* planting a (rustproof!) Russian flag on the ocean floor. Courtesy NOAA.

In addition to video capabilities, each submersible has a set of versatile manipulator arms that a skilled pilot can use to collect biological and geological samples. The arms are also used for many other tasks, such as placing small temperature recorders into hydrothermal vents and gently pushing the sub backward when the pilot wants to avoid stirring up sediments with the thrusters.

As with all manned submersibles, the Mirs are launched and recovered from a support vessel, the *Akademik Mstislav Keldysh*. Since the *Keldysh* operates both Mirs, one submersible can conduct science dives while the other remains in "ready" status in case of an emergency. For example, if the working Mir were to get stuck at the bottom of the ocean, the second Mir could be launched and rushed to the site, where it could work to free the first Mir. Also, some research projects and special tasks such as the lighting of shipwrecks benefit from the use of the Mirs in tandem.

ALVIN

Those who have seen the film *Titanic Revealed* will recognize Alvin (Panel 4.1, Figure 4.2, and Table 4.3) because it was the submersible that Dr. Robert Ballard used to explore the famous wreck. Although it is capable of diving as deep as 4.5 km, Alvin cannot dive autonomously. A support vessel drops the submersible overboard and Alvin begins to sink using a variable ballast system (VBS). To go up or down, the pilot uses the VBS to pump seawater in and out of ballast tanks on the side of the vessel. This elevator-like system has worked for decades, but because of limitations of battery technology, Alvin does not give scientists much time on the sea floor. This means manned deep-sea exploration is an incredibly laborious exercise, akin to exploring Africa with a Jeep!

Panel 4.1. Alvin

Alvin is one of only five deep-diving manned submersibles in the world. The submersible revolutionized deep-sea exploration by providing US scientists with an unprecedented ability to routinely conduct research on the deep-sea floor, at mid-ocean ridges, and hydrothermal vents with exotic marine life now believed to hold clues to the origin of life on Earth.

Over a period of four decades, the submersible has performed more than 4,000 dives and transported 12,000 people to the sea floor to spend 16,000 hr on the bottom to observe and sample the deep. With a current average of 175 dives per year, it has a reliability record greater than 95% over the past 20 years, with the remaining percentage lost mainly to bad weather.

Figure 4.2. Alvin is owned and operated by Woods Hole Oceanographic Institution (WHOI). The long-serving submersible was affectionately named after WHOI engineer, Allyn Vine, whose influence was a key factor in Alvin's development. Operational since 1964, Alvin was the first deep-sea submersible capable of carrying passengers, usually a pilot and two observers. Funded through the National Science Foundation, Alvin has made headlines for locating a hydrogen bomb lost in the Mediterranean Sea in 1966, discovering deep-sea hydrothermal vents in the late 1970s, and exploring the sunken ocean liner, *Titanic*, in 1986. Courtesy NOAA.

Table 4.3. Alvin technical specifications.

Length	7.1 m	Speeds	Cruising: 0.8 km/hr
Beam	2.6 m		Full: 3.4 km/hr
Operating depth	4,500 m	Hatch opening	48.2 cm
Dive duration	6–10 hr	Power	57.6 kWh
Height	3.7 m	Cruising range	5 km submerged at 14 m/min
Draft	2.3 m	Life support	216 man-hours
Gross weight	17 tonnes	Payload	680 kg

Fortunately, new technology is on the way that will enable Alvin's successor to dive deeper and longer. The replacement human-occupied vehicle (RHOV) that will replace Alvin was the outcome of a 2004 National Research Council (NRC) report, *Future Needs of Deep Submergence Science*, which recommended construction of a new, more capable manned submersible (Figure 4.3).

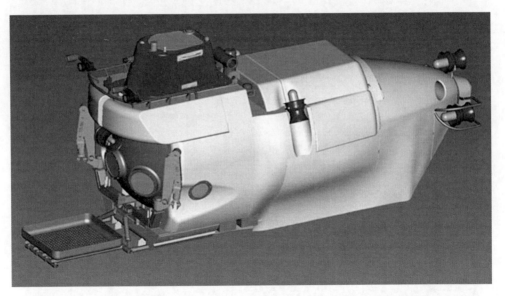

Figure 4.3. In August 2007, Woods Hole Oceanographic Institution (WHOI) awarded Lockheed Martin a $2.8 million contract for the initial design of the Replacement Human Occupied Vehicle (RHOV), which is intended to replace the DSV Alvin. The enhanced design of the RHOV will provide additional space in the vehicle's personnel sphere for its complement of two scientists and a pilot, accommodate a larger science payload, provide improved visibility, and dive as deep as 6,500 m. Courtesy NOAA.

Alvin's successor will include several improvements (Table 4.4), including an increased maximum-depth capability of 6.5 km, a larger personnel sphere, improved speed and maneuverability, and faster descent/ascent times. These capabilities will be realized as a result of overcoming a number of technical challenges. For example, a new VBS will be designed, new types of batteries will be used to power the vehicle,

Table 4.4. RHOV Alvin technical specifications.

Depth	6,500 m	Maximum speed (lateral)	0.5 knots
Sphere volume	4.84 cu m	Maximum speed (vertical)	48 m/min
External science payload	182 kg	Trim angle	$\pm 15°$
Maximum speed (forward)	3 knots		

and new flotation material capable of withstanding greater pressures will be fabricated.

Once the Mark II Alvin RHOV is ready in 2015, scientists and operators will be able to take advantage of a vehicle that will provide a significant capability to not only support deep ocean science, but also to extend the diving envelope of man underwater, as the vehicle will become the deepest-diving manned submersible in the ocean. In common with hardsuit divers, operators of manned submersibles must rely on systems similar to those that sustain astronauts during spacewalks. Another common feature is the harsh environment in which those piloting manned submersibles must operate. Given the extremes of operating a multi-tonne vehicle in such a hostile environment, it is not surprising that those selected as pilots must undergo some very specific and demanding training.

The ocean's astronauts

In 1965, Bill Rainnie and Marvin McCamis climbed inside Alvin and guided it to a depth of almost 2,000 m. It was an epic dive that certified them as the first pilots of the world's deepest-diving research submarine. But, unlike the exploits of fellow explorers, Neil Armstrong and Buzz Aldrin, who rocketed to the Moon 4 years later, Bill Rainnie and Marvin McCamis never became household names. In the years that followed, Alvin pilots quietly piloted the submersible that the *New York Times* described as "a curiously shaped midget submarine, that somewhat resembles a chewed-off cigar with a helmet", to depths up to 4,500 m. While some 70 astronauts have piloted the Space Shuttle, the job of driving Alvin has gone to just 34 men and one woman – an elite group of adventurers who have demonstrated over and over again that they had the "right stuff" to work as drivers of deep-sea vehicles. Thanks to their skills, scientists have had the opportunity to examine previously unreported ocean life, map undersea volcanoes, gather biological samples, and witness sights that seem almost extraterrestrial in origin.

Being certified as an Alvin pilot is no easy task. For every 40 applications to the Alvin pilot training program, only one is accepted. Many wannabe Alvin pilots have engineering degrees, which is a useful qualification, since pilot certification requires drawing by memory dozens of the submarine's complicated ballast, electrical, and hydraulic systems. Since every launch and recovery requires assistance in the water, each pilot must also be a proficient swimmer, and because there are no underwater garages, they also serve as the submarine's onsite mechanic.

Before the prospective pilot-in-training (PIT) is offered the opportunity to train as an Alvin pilot, they must first undergo an informal evaluation period, which involves helping out as much as possible and learning how day-to-day operations work. The evaluation period is an opportunity not only for the Alvin group to assess the budding PIT, but also for the prospective pilot candidate to evaluate the lifestyle and features of one of the most unusual jobs on the planet. Assuming the prospective Alvin pilot thinks the lifestyle is for him/her, and assuming the Alvin vets think the same, the PIT course can start.

One of the first steps in becoming qualified is to learn by rote the checklists of systems, of which there are more than 200. The student pilot has to be able to draw a mechanical diagram of every system from memory – everything from the air revitalization system (ARS) to the VBS. Once the student pilot thinks he/she has mastered the systems, he/she is evaluated by scientists and qualified pilots. The evaluation, which includes four oral examinations lasting between 4 and 5 hr, requires the student to explain how each system works, why it works, and how it is maintained. The first evaluation is conducted by scientists, who ask the student pilot questions about safety and how to complete scientific dive objectives. After the scientists have grilled the student, it is the turn of the veteran Alvin pilots, who ask the candidate to stand in front of a whiteboard and draw engineering systems at random. The second day begins with an oral examination conducted by the deep-submergence engineers at Woods Hole. The engineers have an encyclopedic technical knowledge of Alvin and expect the candidate to have a thorough understanding of what makes the submarine tick. After surviving the technical inquisition by the engineers, the candidate has one more hoop to jump through: the Navy Review Board. For this final phase of assessment, the candidate is sent to the Navy's deep-submergence facility in San Diego, where he/she sits down in front of an admiral and a group of submarine captains. If the candidate is successful at this stage, he/she is awarded the coveted Navy deep-submergence dolphins and cleared to dive. Solo.

Concept of operations

A typical day at sea begins before dawn with a check of Alvin's systems. The pilot and support personnel test all the electronic equipment and ensure the cameras and video recorders are loaded with tape. Once the routine checks have been completed, ballast weights are loaded onto each side of the submersible to help Alvin sink to the sea floor. The mission begins when the scientists step into the submersible and take their positions behind the pilot. Using the VBS, the pilot adjusts Alvin's buoyancy, enabling vertical travel to the research site. In addition to the VBS, the pilot uses the trim system to adjust the pitch angle. Informing the pilot of his/her progress is a temperature-compensated quartz oscillator pressure transducer that provides depth measurements accurate to one-tenth of a meter. As Alvin descends, the pilot maintains radio contact with *Atlantis*, the surface-based support vessel, via an underwater telephone. Simultaneously, the pilot navigates Alvin using a surface tracking and vectoring system in which the submersible emits an acoustic pulse every

3 sec. The Nautronix acoustic system mounted on the support ship receives the pulse and calculates the position of Alvin relative to the ship using the bearing, depression angle, and depth. This information is then integrated with the GPS navigation to log the submersible's geographic position.

Once at the research site, the scientists go to work, making use of Alvin's myriad scientific subsystems. One of the most utilized items of equipment are the cameras that allow researchers to take pictures of features and creatures during the dive and also to document sampling sites to aid post-dive debriefing. To ensure observers achieve good-quality photos, the submersible is fitted with powerful 400-W thallium iodide arc lamps and to permit tactile investigation of features, operators use jettisonable hydraulically powered manipulators. Just like a human, the manipulators are capable of shoulder pitch and yaw, elbow pitch, and wrist pitch and rotate. At the end of the manipulators are "jaws" that are the functional equivalent of opposing overlapping finger pairs, designed to grip instruments and pick up samples.

Armed with lighter batteries and stronger materials, the new RHOV will continue where the old Alvin left off, by operating as a workhorse of deep-sea research, only this time, it will do it as the undisputed deepest-diving manned submersible – a designation it will share with one other vehicle.

SHINKAI

No part of Japan is more than 120 km from the sea. With its shipyards some of the world's most innovative and its Emperor a marine biologist, Japan has quietly honed its technological skills to become a world leader in probing the depths of the world's oceans. Smaller than California and bereft of many natural resources, Japan sees the ocean depths as a repository of biological and mineral riches and is making preparations to mine them both. It is also exploring the abyss to fathom its own geophysical fate, which is tied to nearby oceanic plates that grind past one another, generating earthquakes and tsunamis. The Japanese hope many of these objectives will be achieved through the use of the Shinkai 6500 (Figure 4.4), the world's deepest-diving manned submersible.

Shinkai 6500

The Shinkai 6500 (Table 4.5) represents the sharp end of a comprehensive and ambitious national undersea strategy that began in the early 1970s when the Japanese founded the Japan Marine Science and Technology Center (JAMSTEC) and turned to the West for insights into how to probe the deep. Unsurprisingly, they turned to Woods Hole, which had invented a certain deep-maneuvering submersible called Alvin. Japanese engineers visited Woods Hole in 1973 and inspected Alvin, and following much experimentation, the Japanese launched the Shinkai 2000 in 1981. Nine years later, the Japanese completed the Shinkai 6500, which can carry three people down to 6,500 m.

Figure 4.4. The Shinkai 6500, which made its 1,000th dive on March 15th, 2007, is Japan's deepest-diving submersible. In Japanese, *Shinkai* translates as *deep sea*. For a submersible capable of diving to a depth of 6,500 m, the name is a fitting one because, with its depth capability, the Shinkai 6500 is able to survey approximately 98% of the world's ocean floor. Courtesy JAMSTEC.

Table 4.5. Shinkai 6500 technical specifications.

Physical characteristics

Length	9.5 m	Breadth	2.7 m
Height	3.2 m	Weight	25.8 tonnes

Performance characteristics

Operating depth	6,500 m	Complement	Two pilots, one scientist
Life support	129 hr	Maximum speed	2.5 knots

Notable milestones
- First to discover a rift on the surface of the Pacific Plate at a depth of 6,270 m
- First to film a swarm of deep-sea cold-seep clams at a depth of 6,374 m
- Conducted more than 1,000 dives around the world
- Carried more than 800 scientists

Twenty years later, the gleaming white submersible dives almost every day, helping the Japanese map cracks in the ocean floor, monitoring active geological faults, searching for seabed lava flows, and gathering valuable data that may lead to mineral-resource exploitation. Recently, Shinkai 6500 has been used to retrieve unique bacteria and organisms from the ocean, with the intention of making new drugs (Panel 4.2) and tools for genetic engineering.

Panel 4.2. Ocean exploitation

Biotechnologists are continually searching for microorganisms that can function in extreme environments to enable creation of new processes and industries. For example, acidophiles and alkaliphiles have been studied extensively, and some are used in industrial-scale processes. Additionally, unique enzymes that have been isolated from thermophiles are now available as reagents. Barophiles have also been isolated relatively recently, and while no biotechnological applications have been reported as yet, future applications are anticipated. Organic solvent-tolerant microorganisms have now been isolated all over the world and have considerable potential for application in biochemical engineering. Without doubt, novel extremophiles will bring many benefits to biotechnological applications, which is why the Japanese continue to use the Shinkai 6500 to search for samples.

The inhabitable space of the Shinkai 6500 is contained within a 2-m-diameter pressure hull that can accommodate two pilots and one scientist. While this may sound like a small space, the habitable area is actually a lot smaller due to the myriad instruments installed in the pressure hull. It is the pressure hull that is the key to the Shinkai 6500's operation, since the pressure at 6,500 m exerts a force of several tonnes per square centimeter. At such a pressure, even a slight warping of the hull can lead to structural failure. To guard against such a possibility, the 73.5-mm-thick hull was constructed from a super-strong, lightweight titanium alloy that was machined to a sphere with discrepancies of less than 0.5 mm in diameter at any point on its surface. To help it ascend, the Shinkai 6500 utilizes buoyancy material made of syntactic foam – a material produced by embedding hollow glass spheres into high-strength epoxy resin. It is a process that guarantees sufficient strength and buoyancy even at the higher water pressure.

Helping scientists do their job is a suite of instruments such as acoustic imaging sonar (AIS) capable of showing frontal and sectional images of an underwater object and constructing three-dimensional images. Dexterity is provided by a couple of manipulators, the right having seven degrees of freedom and the left having five degrees of freedom.

In common with the Mir and Alvin, the Shinkai 6500 is reliant upon a support vessel, in this case either the *Yokosuka* or the *Natsushima*, two purpose-built support vessels that deploy, track, and recover the submersible in addition to providing scientists with laboratories.

THE FUTURE OF MANNED SUBMERSIBLES

With only five manned submersibles in existence, some may be wondering whether these vehicles are heading the way of the dodo and whether there is any future for such a highly specialized system. It is a question open to interesting speculation because other than the revamped Alvin and China's new manned submersible (Harmony, which is due to be operational in 2010), there has been little activity in the deep-submersible field in recent years. Although there are several technical challenges involved in sending submersibles to depths exceeding 6 km, one reason for the dearth of new submersibles is simply public perception. Apart from a few publicized documentaries of celebrities such as James Cameron visiting *Titanic* onboard the Mir, the only perceptible public awareness of manned submersibles is when they conduct scientific research. Given the low media profile of this activity, the resultant public perception of manned ocean exploration is that it is almost non-existent, which means there is an absence of political influence for the funding of underwater research. To increase the profile of manned submersibles, those in the industry need to take a leaf out of the manned spaceflight book and promote underwater exploration. If you were to ask the person in the street what they thought the future of the manned space program was, they would probably mention Mars or the International Space Station (ISS). However, if you were to ask the same person what they thought the future of manned underwater exploration was, the chances are they would just give you a blank look!

Another reason for the dearth of manned underwater projects is the cost, which, in common with space exploration, will continue to increase. A big part of the cost problem is that in most situations, manned submersible utilization fails the basic cost–benefit economic test, in large part because of the expense of the surface support ship. One solution to the funding problem is to learn to do more with existing funding, while another remedy may be to develop new technologies. For example, it is possible that with the advent of usable air independent propulsion (AIP) systems such as polymer electrolyte membrane (PEM), it will be possible to develop cost-effective and capable alternatives. By utilizing AIP (Panel 4.3), it will be possible to design small to mid-sized submersibles capable of transiting to the dive site while on the surface before spending days operating underwater, then returning to the surface for the trip home. AIP capability would not only dramatically expand the undersea range and endurance of submersibles, but such a capability would obviate the requirement for a surface support ship, thus cutting the overall system operating costs by as much as half.

Panel 4.3. Polymer electrolyte membrane

A fuel cell is an electrochemical conversion device that combines hydrogen and oxygen to produce water and electricity. Many authorities believe that fuel cells offer the best potential for developing more capable AIP systems. There are several potential fuel cell configurations, but for submersible propulsion, PEM fuel cells have attracted the most attention due to their low operating temperatures (80°C) and relatively little waste heat.

In a PEM device, pressurized hydrogen enters the cell on the anode side, where a platinum catalyst decomposes each pair of molecules into four positively charged hydrogen ions and four free electrons. The electrons depart the anode into the external circuit as an electric current while, on the cathode side, each oxygen molecule is catalytically dissociated into separate atoms, using the electrons flowing back from the external circuit to complete their outer electron "shells". The polymer membrane that separates anode and cathode is impervious to electrons, but allows the positively charged hydrogen ions to migrate through the cell toward the negatively charged cathode, where they combine with the oxygen atoms to form water, which is the only exhaust product.

Making investments in innovative technologies such as AIP and incorporating these into current and future manned submersibles will expand platform capabilities and extend useful lifetime, thereby reducing the pressure to build new systems in the near future. For example, improved multifunction controls could be developed that would support more integrated operations by the pilot. Such an approach has been pioneered by the aerospace industry and significant benefits could be observed if a more holistic methodology was applied to the manned submersible workspace.

Ultimately, manned submersibles will survive because they provide direct access, which offers better information, and allows observers to prioritize plans and ensure efficient operations. But perhaps the strongest argument for manned submersibles is that there is no replacement for in-situ three-dimensional visualization and situational awareness. The very best ROVs in the world cannot duplicate human sensing and visual data processing, nor can they extract directly relevant task information or evaluate and assess relationships and interactions in the same way as a human can. Let's face it, innumerable scientific discoveries would not have been possible were it not for the observations of scientists onboard Alvin and the Shinkai 6500. Just ask those working at Woods Hole or onboard the *Yokosuka*. While some ROV proponents may argue that manned submersibles suffer from viewport constraints due to vehicle design, strategic placement of virtual, augmented reality, telepresence and panoramic imaging systems can more than compensate for these shortcomings.

While manned hydrospace may be considered by some to be a vestige of the past, countries such as the US, Japan, and China continue to pioneer deep-sea exploration, and do so using manned submersibles. Unfortunately, these vehicles still operate in much the same way as they did five decades ago, dependent on the support of an expensive mother ship. For there to be a meaningful role for the future fleet of manned submersibles, they must not only be cost-effective (which means cutting the umbilical cord with the mother ship), but also be low-maintenance, affordable, and capable of achieving what has long been the Holy Grail of manned submersibles: full ocean depth. Fortunately, this new class of submersibles is already being developed by a small group of undersea entrepreneurs and engineers, who preach what appear to be far-fetched ideas with the fervor of evangelists. Collectively, they are in the vanguard of a new dawn for the manned submersible, poised to take their place in the future of sub-sea operations and possibly transform how humans venture into the deep ocean in ways most can scarcely imagine.

5

Personal Submersibles and Underwater Flight

"Are you ready for the other side of the surface? With our solutions you can explore the underwater world – places where no one has been before, with safe and comfortable private submarines. Be the adventurous forerunner and fulfill your underwater dream with your special someone or group of friends. Find privacy and peace in the abyss – whenever, wherever and as long as you desire."

Advertisement by Kemp Marine in response to the
increasing demand of private submarines

The murky gray water rises steadily. Within seconds, it swirls around the feet of the three occupants, but they don't as much as flinch thanks to being ensconced inside a watertight acrylic sphere attached to the front of a canary-yellow submersible, which is now almost completely submerged. As water laps over the occupants' heads, the submersible gradually pitches nose down and the aquanauts get their first glimpse of an underwater world usually reserved for scuba-divers. In the shadow of a seamount rich with corals, the submersible descends slowly as schooling hammerhead sharks cruise effortlessly by. As the strange-looking undersea vehicle continues its descent into the abyss, a family of groupers approaches for a closer look. From their vantage point inside the sphere, the occupants witness the groupers from a distance of less than 2 m, but the encounter is fleeting. The cobalt-blue water quickly turns to complete blackness as the submersible follows a vertical wall and the groupers continue on their way. A glance at the control panel shows the altimeter scrolling past 200 m. As the submersible's thallium iodide light pierces downward, it startles a shoal of mobula rays. For a few moments, they swim a little closer and investigate the source of the light until it becomes apparent that the intruder is alien to their world and therefore potentially dangerous. They break away before the swath of light can find them. Downward the submersible glides, a pair of multi-jointed manipulators mounted above the acrylic sphere and two barrel-shaped thrusters positioned either side of the fuselage. The altimeter scrolls past 400 m. At this depth there is no color – even within close proximity of the iodide lights, it is impossible to tell that the submersible is painted yellow. With one eye on the control panel and the other peering through the acrylic, the pilot gently eases the submersible onto a plateau as the high-pitched pings of the active sonar inform him of the proximity of the ground beneath him. With a yank of the throttle bar, the vehicle stirs up a cloud of silt as the skids connect with the sea floor. Touchdown. The pilot, who also happens to be the owner of the compact undersea vehicle, pulls out his camera, a signal for the other two occupants to do the same.

PERSONAL SUBMERSIBLES

The experience described is presently available only to those with deep pockets. However, with 2,300 mega-yachts operational around the world, some costing in excess of $150 million, the stakes in the game of one-upmanship are rising, and some yacht owners like the idea of having a larger and more unique toy. Such a toy is the personal submersible (PSub).

Most PSubs are underwater vehicles that provide the occupants with a sealed environment equal to that of the surface in terms of pressure, oxygen, and temperature – a capability achieved by a sophisticated life support system (LSS). Another advanced design feature is the 20-cm-thick acrylic sphere that protects the occupants from the extraordinary pressures found in the deep ocean. For example, SEAmagine's Ocean Pearl submersible (Figure 5.1) is capable of diving to 900 m – a depth at which the pressure is 91 times that at the surface.

Figure 5.1. SEAmagine's Ocean Pearl, a two-person submersible, is an emerging type of underwater craft that is becoming increasingly popular among explorers and entrepreneurs. Courtesy Will Kohnen, President/CEO SEAmagine Hydrospace Corporation/SEAmagine.

In addition to a LSS and a strong protective sphere, a typical PSub design shares many features common with conventional submarines. For example, the pressure hull resists the hydrostatic forces imposed by seawater and isolates the occupants from the external environment. Another example is the ballast system. To descend and ascend, a PSub uses a ballast and trim system comprising a main ballast tank (MBT) composed of groups of port and starboard tanks, which are usually vented to seawater at the base. The purpose of the MBTs is to provide the vehicle with the necessary freeboard, stability, and buoyancy while in the surfaced condition. MBTs can also be blown at depth with high-pressure air in an emergency, resulting in a rapid, uncontrolled ascent. To achieve neutral buoyancy, the PSub utilizes a variable water ballast tank (VBT) system (sometimes referred to as hard ballast), the capacity of which is equivalent to the weight of the rated maximum number of passengers. The VBTs are 1-atmosphere, pressure-resistant tanks, often located in the center of the port and starboard MBTs. Water free floods into the VBT tanks when the requisite valves are actuated, and the tanks are emptied by the introduction of pressurized air. Differential longitudinal trim is achieved simply by filling or venting either the bow tanks or the stern tanks to compensate for passenger movement within the pressure hull.

SEAmagine

William Kohnen is the president of an international corporation. You may not think there is anything unusual about that, but the business Mr. Kohnen oversees is dedicated to the design and construction of submersibles. Based in California, the SEAmagine Hydrospace Corporation has been building submersibles since 1995, which makes the company one of the most experienced businesses of its kind. During the 15 years it has been in business, Kohnen's company has developed a small fleet of submersibles that have performed more than 10,000 dives.

The key design concept of SEAmagine's submersibles consists of a self-propelled electric vessel capable of carrying two or three occupants in a 1-atmosphere acrylic cabin to depths of 450–900 m. Climbing onboard one of SEAmagine's submersibles is easy thanks to a cabin that opens as a clam shell. Once inside, the occupants enjoy a shirt-sleeve environment and breathe air at normal atmospheric pressure thanks to a LSS that scrubs the carbon dioxide and replenishes the oxygen. In case you are wondering how safe PSubs are, you need not worry because the submersibles are built to the highest-quality standards in the submersible industry and are classed +A1 by the American Bureau of Shipping (ABS).

Ocean Pearl

One of SEAmagine's most successful submersibles is the Ocean Pearl (Table 5.1 and Figure 5.2a–c). Intended as a recreational sub to be carried aboard large luxury yachts, it features a spherical cabin providing a 360° view for the pilot and passenger, is capable of diving to more than 150 m, has a range of 12 nautical miles, and can conduct several dives a day.

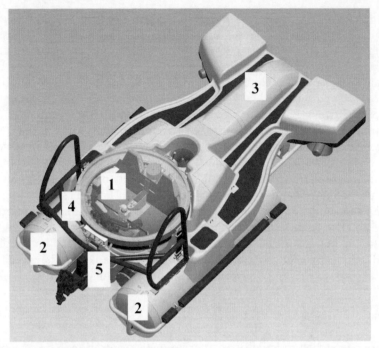

Figure 5.2a. Technical features of the Ocean Pearl. (1) Acrylic cabin. This provides a 1-atmosphere environment and consists of two identical acrylic hemispheres mounted onto upper and lower aluminum rings. The cabin rings are hinged at the front of the vessel and the upper hemisphere tilts open like a clam shell. (2) Port and starboard flotation bladders. The bladders enable the vessel to float high above the water line. Each bladder compartment has a backup reserve bladder that can be inflated in the event of a main bladder malfunction. (3) Tail buoyancy foam. To ensure the submersible remains horizontal, the tail section is filled with foam. (4) Forward-looking sonar. This is mounted at the front of the submersible and is used to avoid obstacles. (5) Drop weight assembly. Ocean Pearl is equipped with 100 kg of ballast that can be released mechanically from inside the cabin in the event of loss of buoyancy underwater. Courtesy Will Kohnen, President/CEO SEAmagine Hydrospace Corporation/SEAmagine.

Special foam in the tail section of the vessel provides buoyancy that keeps the craft in a horizontal pitch on the surface and underwater and in the unlikely event of loss of buoyancy, the submersible is equipped with a drop weight assembly (DWA). The DWA, which is mechanically activated from inside the cabin, is a 100-kg weight that represents the loss of buoyancy that would occur in case the largest floodable volume was compromised. To submerge, the Ocean Pearl relies on a lead ballast system (LBS) comprising lead blocks located underneath the flotation bladder cylinders. The LBS also ensures proper pitch when underwater. Once underwater, the submersible pilot navigates using the forward-looking sonar, which presents information on the submersible's computer monitor. By making control inputs on a joystick, the pilot can operate the port and starboard thrusters, making the submersible go forwards or backwards or rotate on the spot.

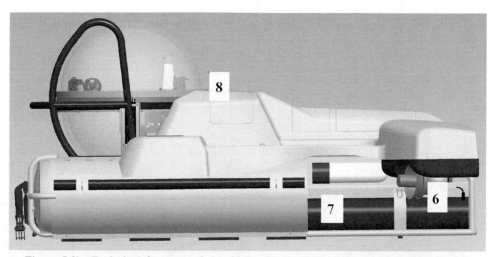

Figure 5.2b. Technical features of the Ocean Pearl. (6) Port and starboard thrusters. These are positioned aft and are controlled by a joystick. The push or pull of the thrusters allow the Ocean Pearl to move forward, reverse, or to simply stay in one place. (7) Permanent lead ballast. These lead blocks located underneath the flotation bladder cylinders allow the submersible to submerge and are positioned to ensure proper pitch underwater. (8) Transponder. The transponder is the device that emits the characteristic acoustic "ping" underwater. The "ping" is detected by a surface-based transceiver and ensures accurate location of the submersible at all times. Courtesy Will Kohnen, President/CEO SEAmagine Hydrospace Corporation/SEAmagine.

Table 5.1. Ocean Pearl technical specifications.

General specifications	
Number of occupants	Two (one pilot and one passenger)
Cabin environment	1 atmosphere
Length	4.53 m
Width	2.44 m
Height	2.34 m
Weight	3,200 kg
Price	$1.195 million
Performance	
Maximum depth	153 m
Propulsion duration	6 hr
Reserve cabin power	> 72 hr
Battery recharge time	6 hr

Communication with the surface is achieved using an ultra short baseline (USBL) transponder, which sends an acoustic ping to a surface transceiver mounted topside, making it possible to track the submersible at all times.

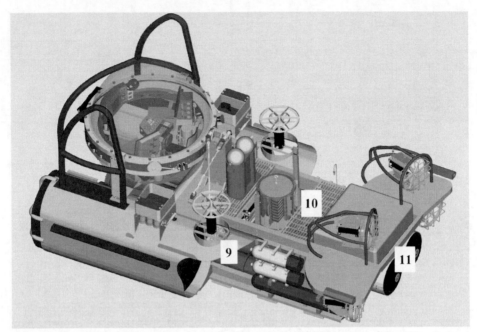

Figure 5.2c. Technical features of the Ocean Pearl. (9) Air ballast tanks. The Ocean Pearl is equipped with two scuba tanks located each side of the vessel. The tanks are used for re-inflating the flotation bladders that lift the submersible back to the surface. The tanks are also used to adjust the vessel's buoyancy while underwater. (10) Electronic pod. This feature contains the entire submersible's electronics and serves as the primary input/output port for the external devices controlled from inside the cabin. (11) Battery pods. These dry pressure vessels contain all the batteries of the submersible and are routed to the Electronic Pod. Courtesy Will Kohnen, President/CEO SEAmagine Hydrospace Corporation/SEAmagine.

Maneuvering the submersible up or down is achieved thanks to a vertical thruster located at the vessel's center of gravity. The thruster is used in conjunction with air ballast tanks (ABTs), which are used for re-inflating the flotation bladders (port and starboard) that lift the vessel back to surface mode. The ABTs are also used for the Buoyancy Control Device (BCD) that allows the pilot to adjust the Ocean Pearl's buoyancy while underwater. Another useful feature is an external diver station that allows the operator to pilot the craft from the exterior and two untrained occupants to be in the cabin. This is particularly helpful for teaching new pilots how to fly the vehicle.

Triumph

For those wishing to venture deeper than 150 m, there is the Triumph submersible (Figure 5.3 and Table 5.2), a self-propelled electric vessel capable of carrying three occupants in a 1-atmosphere acrylic cabin to depths of more than 450 m. In common

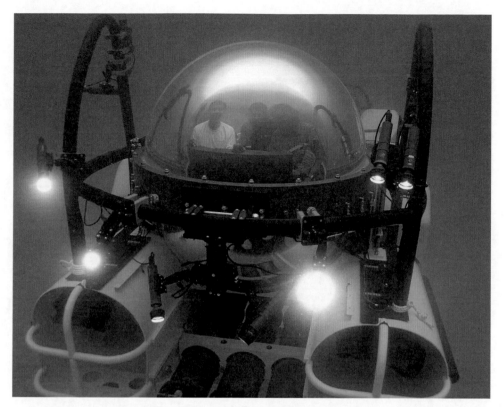

Figure 5.3. Triumph submersible. SEAmagine's largest submersible can carry three occupants to depths of 914 m. Thanks to a spacious 1.5-m acrylic cabin, occupants are afforded a 360° view of their surroundings. A custom-built version of the Triumph – the DeepSee – was built for the Undersea Hunter Group, based in Costa Rica, where tourists can book dives as deep as 457 m. Courtesy Undersea Hunter Group.

with the Ocean Pearl, the Triumph's cabin is equipped with a LSS, which removes carbon dioxide, replenishes oxygen at a controlled flow, and monitors the environment using atmospheric analyzers.

While underwater, the submersible maintains horizontal stability and does not roll or pitch thanks to a static buoyancy and balance configuration. The vehicle's electric propulsion thrusters are fixed and either push or pull, depending on the direction of rotation of the propellers, while depth is controlled by vertical thrusters that are operated by simply using a joystick. Thanks to the Triumph's spacious acrylic cabin, occupants are afforded an exceptional field of view in all directions, which makes the vehicle ideal for undersea tourists and scientists alike.

If you are interested in piloting one of SEAmagine's submersibles, you can book a dive onboard the DeepSee, which was built for the Undersea Hunter Group, which rents its submersible to adventure travelers and as a filmmaking observation vehicle. The submersible has a dive autonomy of 6 hr at a speed of up to 1.7 knots but can remain submerged and operational for as long as 72 hr with its onboard LSS reserve and backup power supply. For those interested in taking an underwater trip on the DeepSee, the prices are listed in Panel 5.1.

Panel 5.1. Booking a dive on DeepSee

1. Dive to a maximum depth of 100 m: $595
2. Dive to a maximum depth of 215 m: $1,095
3. Dive to a maximum depth of 305 m: $1,595
4. Dive to a maximum depth of 457 m: $2,495

Table 5.2. Triumph technical specifications.

Number of occupants			
One pilot		Two passengers	
Passenger environment			
Constant dry	1 atmosphere		
Physical characteristics			
Length	5.33 m	Width	3.11 m
Height	3.05 m	Weight	6.8 tonnes
Life support systems			
Mission air supply	6 hr	Air filtering	Carbon dioxide scrubber
Reserve air supply	72 hr	Emergency supply duration	2 hr
Performance characteristics			
Propulsion type	Electric thrusters (eight)	Propulsion power supply	108 V DC
Propulsion power duration	6–8 hr	Reserve power duration for essential systems	72 hr
Maximum operating depth		475 m (optional upgrade: 914 m)	
Maximum speed		1.7 knots	
Communication system			
Underwater transmitter		OTS Bell 200 25 Khz SSB acoustic transceiver	
Navigation system			
Forward-looking sonar for obstacle avoidance and navigation		Doppler acoustic tracking and navigation system	
Price		$2.85 million	

U-Boat Worx

U-Boat Worx is a Dutch company dedicated to its vision to create "the safest, most reliable, and innovative one-atmospheric submersibles that allow people to explore the underwater world". The company manufactures a single and two-seater version of a recreational submersible called the C-Quester (Figure 5.4). With a price in the region of an expensive sports car, the C-Questers have a top speed of more than 5 km/hr, are safe to a depth of 50 m and provide occupants with a dive time exceeding 2.5 hr (Table 5.3). At a little over 3 m long and weighing just over 1 tonne, the submersible is small enough to fit on a trailer to the nearest boat ramp or be launched from a yacht.

Figure 5.4. Uboatworx's C-Quester is a dedicated super-yacht submersible based on a double-hull configuration known as the Underwater Boat Concept. The first hull is the exostructure that ensures excellent stability, while the second hull is the actual pressure vessel, constructed of high-grade carbon steel. The configuration allows the submersible to perform as a normal boat on the surface, while still retaining underwater maneuverability. Courtesy UboatWorx.

Table 5.3. C-Quester 1 technical specifications.

Physical characteristics			
Length	278 cm	Height	183 cm
Weight	1,030 kg	Drop weight ballast	50 kg
Performance characteristics			
Maximum payload	100/200 kg	Dive time	2.5 hr
Maximum depth	50 m	Maximum speed	3 knots
Life support			
Air reserve	36 hr	Air filter	CO_2 scrubbers
Survival system	36 hr	Monitoring	$\%O_2$, $\%CO_2$
Cost		$105.000	

In common with many of the PSubs currently being developed, entry is jet-fighter style through the submersible's canopy, while steering is performed using a joystick similar to one used in SEAmagine's submersibles. Given the similarities between the C-Quester and a jet, it is not surprising that the experience is not that different from flying an aircraft, although it is a lot quieter thanks to the three electric motors that supply the motive force.

HYDROBATIC SUBMERSIBLES

Comparatively affordable and light enough to launch from a yacht, the submersibles designed by companies such as SEAmagine and U-Boat Worx have opened the depths to a growing army of private explorers. While this is a boon for the new generation of underwater tourists, there are some for whom 457 m is not deep enough, while, for others, scooting around on the ocean floor at 3 or 4 knots is just not fast enough. While the designs of the Triumph and C-Quester submersibles are clearly cutting-edge, the buoyancy/ballasting principles by which they operate are still the same as those used by manned submersibles more than five decades ago. Although they might be freed from the constraints of a mother ship, the vehicles are still slow to maneuver and are limited to relatively shallow depths; 457 m is still a long way from realizing the Holy Grail of full ocean depth. In the eyes of one pioneer inventor, what was needed was a quantum leap forward in submersible design. That inventor was Graham Hawkes.

Hawkes Ocean Technologies

He has been described as a mad scientist, but if one man can revolutionize the field of underwater exploration, it is Graham Hawkes. An internationally renowned ocean engineer and inventor engineer, Hawkes has been designing and building deep-sea exploration vessels for more than 20 years. In fact, Hawkes has been responsible for designing a significant percentage of the planet's manned underwater vehicles. Among his many creations are the WASP and Mantis atmospheric diving suits (ADS), which were built for research and industry, while his submersibles such as the Deep Rover have featured in films such as James Cameron's 3-D Imax film, *Aliens of the Deep*.

In addition to designing and building the submersibles, Hawkes also flies them and currently holds the world record for the deepest solo dive (910 m) – a mark set while test-piloting the Deep Rover submersible. However, while the Deep Rover is one of the most advanced submersibles ever built, it is Hawkes' version of a flying submersible that has grabbed the attention of those interested in exploring the depths. Developed by Hawkes Ocean Technologies (HOT), an engineering skunk that creates the world's most advanced deep-sea craft, Deep Flight (Figure 5.5) has introduced an entirely new concept to underwater exploration.

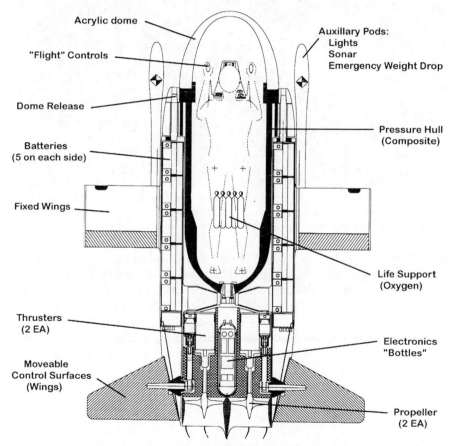

Figure 5.5. Deep Flight I was a prototype submersible designed to reach the deepest part of the ocean. Launched in 1995, the submersible uses the distinctive inverted wings to generate "negative" lift instead of a traditional ballast system to control dive and ascent. Courtesy Hawkes Ocean Technology.

Figure 5.6. The Deep Flight Super Falcon. Those familiar with the cartoon character Tintin may recognize the concept from the 1972 French comic book, *Tintin and the Lake of Sharks.* Courtesy Hawkes Ocean Technology.

Deep Flight I (DF I) was Hawkes' prototype vehicle that set the standard for HOT's radical new approach to submersibles. Instead of using a traditional ballast system to control the descent and ascent, DF I uses inverted wings to generate "negative lift" that pulls the vehicle down and creates a force that counteracts the submersible's buoyancy. While DF I looks like it dropped out of the future, it was merely a test-bed for even more advanced technologies that Hawkes used to develop the Super Falcon, a winged submersible he hopes will revolutionize manned underwater exploration.

Deep Flight Super Falcon

In 2008, Hawkes unveiled the Deep Flight Super Falcon (Figure 5.6), a sleek, winged submersible, which operates by utilizing the principles of lift and drag, much like an aircraft. The vehicle, which provides unprecedented sub-sea speed and maneuverability, will only be flown by those with deep wallets (the vehicle's price tag is $1.5

million) to depths exceeding 400 m. Thanks to its agility, research institutions have expressed interest about the winged vessel's potential for studying and following exotic, mysterious sea creatures, such as super sharks and the ever elusive giant squid.

A fourth-generation winged submersible, the Super Falcon, is the newest vehicle in the Deep Flight series of submersibles built by HOT. Whereas most submersibles operate like an underwater balloon, the winged Super Falcon is capable of true underwater flight, thereby enabling a safer and more efficient means of moving through the water. The resemblance to a fighter jet is not accidental; this submersible is designed to literally fly through water and has about as much in common with traditional submersibles as a balloon has with the F-22 Raptor. The submersible's unique winged design provides the pilot with a no-holds-barred aerobatic flight experience through liquid blue space – a marked contrast with the cumbersome elevator experience of a traditional submersible. Whereas conventional submersibles depend on ballast, a design feature that restricts their use to waters no deeper than their operating depth, the Super Falcon (Table 5.4) is a fixed positive buoyancy (FPB) craft and therefore has no such restrictions, so the sub can never accidentally sink to dangerous depths. In fact, the submersible has 800 kg of built-in positive buoyancy, which is more than sufficient to maintain the craft safely on the surface if required.

Table 5.4. Deep Flight Super Falcon technical specifications.

Physical characteristics			
Length	6.7 m	Weight	1,818 kg
Height	1.6 m	Width (wings deployed)	3.0 m
Performance characteristics			
Maximum payload	Two persons	Dive time	5 hr at 4 knots
Maximum depth	300 m	Maximum speed	6 knots
Maximum descent rate	60 m/min	Maximum ascent rate	120 m/min
Pitch/roll	± 30°	Maximum thrust	231 kg
Emergency systems			
Buoyancy	1,800 lbs surface inflation	Carbon dioxide	Redundant manual scrubber
Oxygen	Redundant manual	Cockpit	Release mechanism

Much like a spacecraft, the Super Falcon is equipped with a redundant cabin environment system that provides the pilot with more than 24 hr of submerged life support. In common with the fighter jet that it resembles, the Super Falcon operates by a fly-by-wire system comprising a series of micro-controllers connected to electromechanical actuators that are, in turn, connected to the submersible's rudders, elevators, and ailerons. The pilot controls pitch, roll, and yaw inputs using the microcontrollers and monitors the Super Falcon's progress through the water by means of a computerized track plot displayed on a heads-up (HUD) flight display, just like a fighter aircraft. While it might sound easy to operate, you cannot just buy a Super Falcon and start flying. Just like flying an aircraft, the Super Falcon requires necessary training.

Underwater flight school

The world's first underwater flight school opened in the Bahamas in February 2003. Open to anyone interested in helping pioneer sub-sea flight, HOT's flight school offers a one-dive experience, a 1-day course, or a 3-day licensing option (Table 5.5). In addition to learning to be one of the few humans qualified to fly underwater, students also have the opportunity to meet scientists, fellow adventurers, and Graham Hawkes. The airborne equivalent would be taking a trip to the Mojave Desert to be trained to fly one of Burt Rutan's (of SpaceShipOne fame) experimental aircraft.

Table 5.5. Levels of flight experience.

Course	Cost	Description
One dive	$2,500	Short land-based orientation and 45–60-min dive in the Super Falcon, with minimal instruction in maneuvers and fundamentals of flight control
1 day	$5,000	Introduction to the Super Falcon, familiarization with controls and instrumentation, pre-dive checklist, safety procedures, introductory shallow orientation dive (15 min), followed by instruction on basic maneuvers and fundamentals of flight control, followed by a longer dive to 30–60 m (45–60 min). The student pilot practices procedures and maneuvers from the rear seat while Graham Hawkes flies in the front seat as Pilot in Command (PIC)
3 day	$15,000	Training and licensing course First day follows the schedule of the 1-day course. Day 2 includes orientation to three-axis underwater flight, use of communications systems, hands-on piloting and practice of thrust control, and three-axis maneuvers. Also includes two longer and deeper dives, each approximately 60 min in duration

For those purchasing a Super Falcon (the first was delivered to venture capitalist, Tom Perkins, in November 2008), it is necessary to enroll into the 3-day course (Table 5.6). Day 1 includes land-based instruction followed by a 45–60-min dive in the Super Falcon with instruction in maneuvers and fundamentals of flight control. Day 2's schedule includes a review of controls and instrumentation before instructing the prospective pilot on necessary safety procedures and pre-dive checklists. This is followed by an orientation to three-axis underwater flight, the use

Table 5.6. Deep-flight training schedule.

Time	Event
	DAY 1
0800	Individual fit and adjustment of: • seat controls • heads-up display Hands-on review of: • controls • life support • power-up • communications Familiarization with cockpit: • dome closure and opening • monitoring/adjustment of life support • surface emergency procedures
1000	Dive 1 (Pilot A): Short shallow orientation from back seat; life support, cockpit closure, communications
1100	Dive 2 (Pilot B): Short shallow (beachside) orientation from back seat; life support, cockpit closure, communications
1200	Lunch
1300	Dive 3 (Pilot A): Short shallow orientation dive to flight controls
1430	Dive 4 (Pilot B): Short shallow orientation dive to flight controls
	DAY 2
0800	Dive 5 (Pilot A): Wet orientation; VFR short training dive (diver depths)
1000	Dive 6 (Pilot B): Wet orientation; VFR short training dive (diver depths)
1200	Lunch
1300	Dive 7 (Pilot A): First dive below diver depth
1430	Dive 8 (Pilot B): First dive below diver depth
	DAY 3
0800	Planning and preparation for final dive (2 hr hatch to hatch)
1000	Dives 9 and 10: Final dives. The goal is that the student is the hands-on pilot for 100% of the 2-hr dive. Graham Hawkes, as Pilot in Command, will coach as needed and will navigate and operate cameras from the rear. The goal is to have the student pilot function on a purposeful dive deeper than 100 m
1200	Lunch
1900	Wrap up. License certification/dinner

of the communication system, and hands-on piloting. Once the student pilot is comfortable with thrust control and three-axis maneuvering, he/she has the opportunity to pilot the Super Falcon while seated in the front cockpit for the first of two 60-min dives. In case the student has any problems, Graham Hawkes acts as Pilot in Command (PIC) in the back seat. Day 3 starts with a review of controls and instrumentation and safety procedures before moving on to more advanced instruction on the subjects of emergency procedures, complex flight maneuvers, and underwater navigation. The student then has the opportunity to practice what he/she learnt in the classroom by flying the Super Falcon while Graham Hawkes acts as PIC in the back seat. The Day 3 dive is a proficiency qualification dive that lasts up to 2 hr and requires the student to demonstrate responsibility for cockpit management, use of checklists, life support, and flight controls. If Hawkes is satisfied the student meets the requirements, certification is awarded and the student receives a license as "crew pilot". This is rather like a regular student aircraft pilot having permission to fly solo. Once the "crew pilot" has accumulated additional flight experience, he/she has the "crew" restriction removed.

Deep Flight II

While the Super Falcon is undoubtedly a versatile, envelope-pushing submersible, Deep Flight II (DF II, Figure 5.7), its successor, may ultimately transform how humans explore the deep ocean in ways that most observers can scarcely imagine today. Looking like some bizarre underwater glider, built with ultra-strong ceramic–metal composites, and propelled by the latest in lithium battery technology, DF II (Table 5.7) will be the most agile and deepest-diving underwater vehicle yet.

Figure 5.7. Deep Flight II. This revolutionary submersible has been designed but has yet to be built, and will be strong enough to dive down to 6,000 m. Courtesy Hawkes Ocean Technologies.

Table 5.7. Deep Flight II technical specifications.

Physical characteristics			
Length	4.8 m	Weight	2,270 kg
Height	1.2 m	Width (wings deployed)	4.5 m

Performance characteristics			
Maximum payload	Two persons	Dive time	8 hr
Maximum depth	11,033 m	Maximum speed	6 knots
Maximum descent rate	120 m/min	Maximum ascent rate	210 m/min
Pitch/roll	$\pm\ 60°$	Maximum thrust	> 200 kg

Derived from Hawkes' experience of building a similar craft for millionaire adventurer, Steve Fossett (Panel 5.2), the DF II will be flown by the pilot lying in an unusual stomach-down position. While such a position may sound uncomfortable, it mimics the posture of scuba-divers and ensures the pilot's eyes are as close as possible to the viewport.

Panel 5.2. Challenger

DF II's predecessor was a similar vehicle called Challenger, which was custom-built for adventurer, Steve Fossett, who planned to dive to the bottom of the Marianas Trench. Following Fossett's death in a plane crash in 2007, ownership of the vehicle passed on to Fossett's widow, and Hawkes turned his attention to applying what he had learnt in designing the DF II.

DF II is as radical it gets in the world of submersible technology, but if development, testing, and financing (DF II will cost an estimated $10 million) go to plan, then, in 2015, underwater exploration will relive its "man on the Moon" moment when Hawkes touches down 11 km below the surface.

Sub Aviator Systems

Given the success of HOT, it is not surprising other companies are using the same technology. Take Sub Aviator Systems (SAS), for example. Their Super Aviator submersible is a radical overhaul of a craft the company bought from Hawkes in 2007. In common with HOT's creations, the Super Aviator (Table 5.8) flies underwater and is piloted using joystick and rudder pedals. It was envisioned and developed by SAS co-founder, John Jo Lewis, a self-confessed underwater addict who gives in to the urge to soar beneath the waves at every opportunity. Lewis's hope for SAS as an enterprise is that it will use underwater flight technology to increase awareness of the ocean environment and of the dire threats to our ocean

Table 5.8. Super Aviator technical specifications.

Physical characteristics			
Length	6.6 m	Weight (launch)	2,225 kg
Height	1.85 m	Width (wings deployed)	3.75 m

Performance characteristics			
Maximum payload	Two persons	Dive time	3 hr
Maximum depth	300 m	Maximum speed	5.2–7.9 knots
Maximum descent rate	95 m/min	Maximum ascent rate	180 m/min
Pitch/roll	$\pm$ 30–60°	Maximum range	20 nautical miles

planet posed by global warming and unsustainable exploitation. Working with him is Phil Nuytten, a pioneer of the modern commercial diving industry, who helped SAS modify Hawkes' original version of the submersible.

The Super Aviator can best be described as a hydrobatic submersible, designed to fully explore underwater flight. Lightweight, high-powered and constructed using a composite airframe with wings, thruster, and flight controls, it not only looks like an aircraft, but it *flies* like an aircraft. Capable of maneuvers normally reserved for aircraft, this revolutionary craft combines the freedom of scuba and depth capability of a submersible with the low intrusiveness of a stealth submarine. Hardly surprising, then, that underwater filmmakers rave about it:

> "I have never seen any craft that allows you to work in the water column, or at depth, like you can with the Super Aviator. As a diver I am ever aware of how awkward and limited we are when trying to work in the ocean. There is simply no easy way to stay at depth without incurring the penalty of decompression. And as an underwater cinematographer who has been very keen on observing and filming marine animals I really think that the Super Aviator might be the bridge that lets us spend more time in their world a tool that gives us a better understanding of what is happening beneath the waves."
>
> Jason Sturgis, underwater filmmaker

Pilot training

Budding underwater pilots have the chance to earn their wings at underwater flight schools, where they have the opportunity to "fly" the Super Aviator. Fitted with several redundant safety features and specialized emergency equipment, the Super Aviator is an ideal submersible in which to learn (Table 5.9) how to operate a PSub down to depths of more than 150 m.

Table 5.9. Sub-Aviator Systems Flight School.

---

1 Day – *Basic Flight Skills* (Course fee: $3,350)
This course prepares the student for their first underwater experience in a winged submersible:

- Pre-flight instruction on safety systems
- Flight control instruction
- One or two dives (total dive time: 1–2 hr)

---

2 Day – *Co-Pilot Flight Control* (Course fee: $6,150)
This course provides instruction on flight operations and is designed to give students "stick time":

- Communication equipment instruction
- Life support equipment familiarization
- Emergency equipment
- Basic and advanced underwater flight maneuvers
- Three or four dives (total dive time: 3–5 hr)

3-Day – *Advanced Flight Competency* (Course fee: $8,650)
This course is designed to allow students to hone their flight skills so they are able to pilot the submersible without an instructor:

- Operation of communication, life support, and emergency equipment
- Extended dive times
- Demonstration of flight skills
- Specialized missions (e.g. wreck inspection)
- Pilot in Command flights
- Five or six dives (total dive time: 5–6 hr)

Not content with the capabilities of the Super Aviator, SAS has partnered with Nuytco Research Ltd of Vancouver to build the next generation of hydrobatic winged submersibles that will be capable of operating to depths of 600 m. With its lean fuselage, powerful thrusters, and cutting-edge technology, the OrcaSub (Figure 5.8), which will sell for $2,190,000, will feature more than 80 hr of life support, underwater communication, and will be even faster and more maneuverable than the Super Aviator.

GAME CHANGE

The depths of the Earth's oceans are largely unexplored territories that contain a wealth of resources just waiting to be tapped. By some estimates, humans have explored only 5% of the ocean and most of that is down to depths of only 300 m. Beyond this depth is a black abyss that plunges to 11,033 m – a region that will soon be within reach of manned submersibles thanks to the efforts of Graham Hawke. But, despite the versatility and maneuverability of these submersibles, human presence will be fleeting and temporary, and while a short-term presence may be just fine for most underwater explorers, there are some who crave something more permanent. It may sound like an impossible dream, but there are visionaries who dream of building mankind's first permanent underwater settlement.

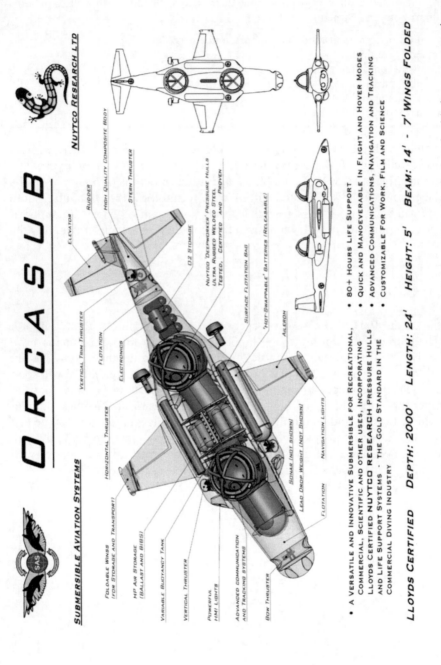

Figure 5.8. The OrcaSub, built by Sub Aviator Systems of Redondo Beach, California. Equipped with long-range underwater communications, advanced sonar capability and a dedicated emergency weight-drop system, the OrcaSub is a state-of-the-art vessel that has a ticket price of $2,190,000. Courtesy Sub Aviator Systems.

6

Ocean Outpost

Humans have learned to adapt to life on almost all parts of the Earth's surface and have pushed the limits of technology to allow the exploration of space. In contrast, the oceans covering over 70% of the planet's surface are relatively unexplored, except for some comparatively shallow excursions made by saturation divers and manned submersibles. However, the prospect of living under the sea has always held a fascination for man and over the last 40 years, a number of underwater habitats have been developed. While none of these has supported a permanent human presence, thanks to developments in sub-sea technology and the efforts of visionaries such as Dennis Chamberland, what was once the subject of science fiction novels may very soon be achieved. While the most lucrative market for permanent underwater habitats will undoubtedly be divers, the dream of living in the ocean will also appeal to adventure-seekers and those wanting to earn their certification as "aquanaut".

An aquanaut is someone who has spent at least 24 continuous hours underwater and the majority of this elite group has been certified in only three underwater habitats. The Jules Undersea Lodge, described here, holds the record, having issued nearly 4,000 certifications, while its underwater neighbor, MarineLab, has certified more than 2,500 aquanauts. The other 500 or so aquanauts were certified in the now defunct Hydrolab, which was operational between 1966 and 1984. While 7,000 certifications sounds like a lot, if you consider these credentials were achieved over a 40-year period, it is not very impressive. Furthermore, less than 5% of those aquanauts stayed beneath the surface for longer than 24 hr and less than 1% stayed for a week or more. In fact, of all those who achieved the status of aquanaut (Panel 6.1), only 25 have ever lived underwater for up to 30 days. It is not a very long time when you consider that the longest-duration spaceflight is 437 days – a record set by Russian cosmonaut, Valeri Polyakov, between January 8th, 1994 and March 22nd, 1995.

Despite there being so many certified aquanauts, most people have little idea what qualifies a person to hold such a title or what an underwater habitat is, so it is worthwhile explaining these terms. First of all, underwater habitats are underwater structures in which people live for extended periods and carry out most human

Panel 6.1. Meet aquanaut, Chris Olstad

Over the past 25 years, Chris Olstad has logged more hours living and working underwater as a professional aquanaut than any human. In addition to his work as a Marine Ecology Field Instructor, he also works as Operations Director for the MarineLab Undersea Laboratory for which he has directed more than 500 scientific and educational missions. Some of these missions have involved the NOAA and NASA, which conducted research investigating the effects of saturation diving. As a diver/scientist, Olstad has also worked on several oceanographic cruises to the Gulf of Mexico using underwater photographic techniques to assess oil rig impact on the nearby coral reef and, as a submersible mechanic, he worked on manned and unmanned systems rated from 3,000 to 6,000 m.

Figure 6.1. Space exploration has traditionally overshadowed Man's fascination with the ocean. Courtesy NASA.

activities, such as working, resting, eating, and sleeping. In this context, "habitat" is used to describe the interior and immediate exterior of the structure but not the surrounding marine environment. An underwater habitat must meet human physiological requirements and provide comfortable environmental conditions such as a suitable physical environment (pressure, temperature, humidity), chemical environment (water, food, waste products), and biological environment (micro-organisms). Given these requirements, it is not surprising that most underwater technology is shared with diving, submarines, and, of course, spacecraft. You may be wondering why anyone would want to spend any length of time underwater, but this is rarely a question that comes up when discussing manned spaceflight (Figure 6.1) – an endeavor in which the challenges of coping in an environment where air is excluded are no greater than those undersea.

Unfortunately, since Man's desire to live in and become part of the ocean has always taken second place to Man's dream to fly, underwater exploration has not made as much progress as space exploration. Nevertheless, there have been several underwater habitats designed, built, and used around the world since the early 1960s, either by private individuals or by government agencies. Many have been used almost exclusively for research that has investigated the physiological processes and limits of breathing gases under pressure, while some advertise their use as underwater resorts. In this section, we take a look at each type of habitat, before peering into the future of underwater outposts and, finally, at the possibility of humans living permanently underwater.

PRESENT-DAY UNDERSEA RESORTS

Jules Verne Undersea Lodge

Located at a depth of 6.5 m near Key Largo, Florida, the Jules Verne Undersea Lodge is the world's first and only undersea habitat, and its name is not just a marketing gimmick, since the entire structure is underwater. Entry to the lodge is via a moon pool entrance that leads to a wet room. Here, guests can leave their diving equipment and enjoy a hot shower before entering the living area. Designed for comfort, the hotel's bedrooms feature all the accessories you would normally expect to find in underwater living quarters. Thanks to an unlimited supply of diving cylinders, guests can explore the underwater world at any time during their stay or simply opt to view marine animals through one of the large viewports. For many guests, their stay in the underwater lodge is an unforgettable experience. For example, one couple decided on a career change and now operates a dive shop, while another couple named their baby after Jules!

The Lodge (Table 6.1) was envisioned by Neil Monney and Ian Koblick (Panel 6.2), who named their creation in honor of Jules Verne, author of *Twenty Thousand Leagues under the Sea*. Koblick, president of the Marine Resources Development Foundation, is author of *Living and Working in the Sea* and is considered a world authority on the subject of living under the ocean, while Monney served as Director of Ocean Engineering at the US Naval Academy and has extensive experience designing underwater habitats.

Six meters is not deep, but because guests stay underwater for several hours, their

Table 6.1. Jules Undersea Lodge.

Specifications

Depth	9 m	Width	6 m
Length	15.1 m	Height	3.3 m
Weight	136,363 kg (dry weight)	Material	Steel and acrylic
Wet room chamber	3 m (base width) × 6 m (length)		
Accommodation	Two private (2.4 × 3 m) bedrooms with 1.06-m viewport. One (2.4 × 6 m) common room with mini kitchen, dining, and entertainment areas. Fitted with 1.06-m viewport.		

Packages

Luxury: Check-in: 1:00 pm. Check-out: 10:00 am. Includes all dive gear, a gourmet dinner, and breakfast. Opportunity to earn the PADI Habitat Specialty Certification – $150.00 extra	$475.00 per person per night
Mini Adventure: 3-hr visit	$125.00 per person
Romantic Getaway: exclusive use of the lodge. Extras include mood music, fresh flowers, seafood appetizer, and chef service for breakfast	$1,295.00 per night, per couple

---

Panel 6.2. Ian Koblick

As an aquanaut, author, explorer, marine consultant, *and* technical advisor, Ian Koblick is one of the foremost authorities on undersea habitation. In the mid 1970s, as president of the Marine Resources Development Foundation (MRDF), he was responsible for envisioning the La Chalupa research lab, the most advanced underwater habitat of its time. As an aquanaut working from the La Chalupa lab, Koblick developed nitrogen–oxygen mixtures that helped divers open up much of the continental shelf to exploration. As president of the MRDF, Koblick also operates the MarineLab research lab and education program in Key Largo, Florida.

---

stay qualifies as a saturation (SAT) dive. Because of this, guests must refrain from flying and diving for 24 hr after surfacing. To prepare guests for their SAT dive, the Lodge's dive instructors present a 3-hr class that describes the equipment and procedures necessary to ensure a safe stay. Once underwater, guests often spend much of their time diving the local reef or visiting MarineLab, an underwater laboratory devoted to research and education. They also avail themselves of the opportunity to earn their Aquanaut Certificate, which is presented to those divers spending 24 hr underwater. To ensure safety of the guests, the habitat is monitored by a Mission Director from a land-based Command Center connected to the lodge by an umbilical cable that delivers fresh air, water, and power.

PRESENT-DAY UNDERSEA RESEARCH HABITATS

MarineLab

Conceived in 1970, and built by midshipmen of the US Naval Academy in 1980, the MarineLab Undersea Laboratory is the world's longest continually operated underwater research habitat. It was developed by Dr. Neil Monney, Director of the Oceaneering Department of the Naval Academy, and has been operational since July 1984. Constructed of steel, the bus-sized cylinder features a large observation port at one end and an acrylic observation sphere mounted beneath. Inside the habitat are three bunks, a microwave oven, refrigerator, and a sink. Separated from the wet room, which houses a shower and portable toilet, are the main living area and laboratory. Air is supplied via a low-pressure compressor by an umbilical from a shore facility that also houses a VHF radio-intercom system, TV monitors, bunks, and a desk for the Operations Director, first-aid supplies, dive logs, and emergency stand-by diving equipment. The umbilical also carries two low-pressure air supply hoses, communications cables, a 110-V AC power supply, 12-V DC power supply, and hot and cold water supply hoses.

Aquarius

While most people know that NASA, together with its international partners, operates the International Space Station (ISS), it may surprise you to learn that the space agency is also involved in the operation of an "inner space" station called Aquarius. Owned by the National Oceanic and Atmospheric Administration (NOAA) and managed by the University of North Carolina at Wilmington (UNCW), Aquarius is the world's only undersea laboratory dedicated to marine science and education. It also happens to be very cheap when compared to the astronomical costs of manned spaceflight. For example, the cost of operating the underwater base is between $1.3 and $1.5 million a year,[1] which translates into an operating cost of about $10,000 per day. While this is a higher day rate than surface-based diving programs, a 10-day Aquarius mission would take more than 60 days if conducted using surface-based technology, and few scientists have the time to spend months in the field when a 10-day Aquarius mission can be used to accomplish the same goals. Sitting in 18 m of water, the facility (Figure 6.2) is a unique ocean

---

[1] By comparison, a single Shuttle mission costs almost half a billion!

Figure 6.2. Aquarius is an underwater laboratory located in a sand patch next to coral reefs in the Florida Keys National Marine Sanctuary, at a depth of 20 m. Inside are all the comforts of home, including bunks, a shower and toilet, hot water, a microwave, a refrigerator, and air conditioning. Courtesy NOAA.

science and diving facility that provides state-of-the-art undersea technology to train students, Navy Divers and astronauts, who regularly conduct 2-week missions onboard.

Since 1993, the lab has supported more than 90 missions and produced over 300 peer-reviewed scientific publications in addition to providing invaluable training to astronauts who use the facility as a test bed for advanced communications that allow for the development of cutting-edge technologies such as telemedicine (Figure 6.3).

Missions are usually conducted on a monthly basis from April through November and prior to each mission, aquanaut trainees undergo 5 days of training. During each mission, a surface-based support crew monitors the aquanauts and habitat from a command center at the shore base and divers are available around the clock for emergencies. By underwater habitat standards, the 85-tonne habitat is surprisingly spacious, providing ample living and working space for a six-person crew. Unlike the Jules Verne Lodge, the purpose of Aquarius is to conduct science, so it is not surprising that much of the habitat's interior is filled with lab equipment, which scientists use to examine the results of underwater investigations.

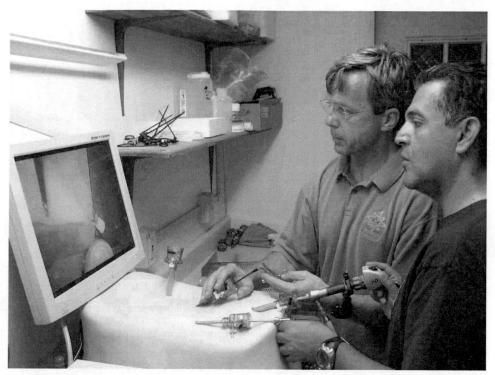

Figure 6.3. Dr. Mehran Anvari (right) assists Canadian astronaut, Robert Thirsk, through a laparoscopic procedure on a training model during a training session for the NASA Extreme Environment Mission Operations (NEEMO) project. Courtesy NASA.

To go to work, Aquarius occupants simply jump into the moon pool. Since the habitat is at ambient pressure, the moon pool can remain open because the equivalent air pressure inside prevents the water from flowing in. Although the habitat's base-plate rests in approximately 19 m of water, Aquarius is mounted off the bottom at a depth of approximately 14 m – a depth known as the "hatch depth". At this depth, non-saturated visitors (i.e. regular scuba-divers) to Aquarius have less than 80 min before they must return to the surface if they want to avoid risking decompression sickness (DCS). However, the Aquarius aquanauts can stay indefinitely, although at the end of the mission, they must undergo 17 hr of decompression, after which they can add the title "aquanaut" to their list of qualifications.

If you want a chance of becoming an aquanaut onboard Aquarius, you should first become a scientist (failing that, you can apply to become an astronaut!) so you can serve as a Principle Investigator (PI). Once you have a research study approved, it is simply a case of submitting the appropriate medical forms to the Mission Coordinator and ensure you comply with the prerequisites (Panel 6.3) at least 2 months prior to the start of your mission.

Panel 6.3. Aquarius mission prerequisites

- Scuba certification by a nationally recognized sport diving agency, NOAA, or military
- Minimum of 50 logged dives
- Passing an American Academy of Underwater Sciences (AAUS) medical examination
- Completion of pre-mission Aquarius training

In addition to the paperwork, you will need to bring your own diving gear, although the program provides buoyancy compensators, cylinders, weights, and regulators. Finally, before being eligible for training, you need to attend a pre-saturation medical exam, which is performed by the Diving Medical Officer (DMO) to determine fitness.

Training to become an aquanaut consists of briefings, pool training, open-water training, and orientation dives to Aquarius. Helping you are several surface-based staff who participate in briefings and accompany you on dives to the habitat. The training is as much an evaluation, since you must demonstrate good diving techniques and a safe attitude to qualify for participation in the program. Common to all the briefings and training in general is an emphasis on safety, which is paramount, since there is a big difference between SAT diving and regular scuba-diving. If there is a problem during a scuba-dive, the surface is always an option, but this is not the case with SAT diving. If you have a problem, you must be capable of solving it in the habitat. It is an aspect of the mission that is hammered home during training (Table 6.2).

Table 6.2. Aquarius training schedule.

Day 1	
0830	Introduction briefing
0900	*Swim evaluation*
	• 400 yards freestyle < 12 min and 10-min tread (no fins, no mask)
	• Underwater swim evaluation: one breath, 25 yards
	• Ditch/don mask and snorkel and clear (pool)
1000	*Dive equipment briefing*
	• Equipment/configuration briefing and harness adjustment
	• Source drill, shut-down drill, safety (s) drill
	• Air sharing/buddy breathing briefing, buddy awareness briefing
1200	*Habitat site orientation briefing*
	Dive 1: Checkout dive (45 feet/45 min)
	• S-drills, buoyancy control, mask clearing, regulator clearing/retrieval
	• Buddy breathing (stationary), air sharing (stationary)
	• Buddy breathing (swimming), air sharing (swimming), shut-down drills

Day 2

0800	Safety and procedures briefing
0900	Excursion emergency procedures briefing
1000	Dive equipment briefing
	• Night dive briefing and excursion/boundary line orientation
	• Line exercise 1, line exercise 2, compass navigation exercise
1300	Dive 2 (50 feet/45 min)
	• S-drills, line exercise 1, shut-down drills
1400	Dive 3 (50 feet/45 min)
	• S-drills, line exercise 2, shut-down drills

Day 3

0800	Food/provisioning briefing
0830	Camera/video housing briefing – effects of pressure differential
0900	Dive equipment briefing
	• Sausage buoy deployment and VHF briefing (exercise/communications check)
1200	Dive 4 (50 feet/45 min)
	• S-drills, lost line exercise, shut-down drills
	• Diver rescue on surface (airway, rescue breathing)
1300	Dive 5 (50 feet/45 min)
	• S-drills, deploy sausage buoy ($\times$ 2), shut-down drills
1500	Fill tanks and rinse gear
1900	Video camera briefing

Day 4

0800	Aquarius orientation briefing
0900	Aquarius emergency procedures briefing
1000	Dive equipment briefing
1200	Dive 6 (60 feet/50 min)
	• S-drills, site orientation, fill tanks and test communications at site fill station
	• Swimming air sharing from site to habitat (mask off)
1300	Dive 7 (Aquarius habitat orientation)
	• S-drills, habitat orientation, Aquarius emergency procedures
	• Habitat exterior orientation
1500	Physical review
1600	Excursion tables
1700	Chamber briefing
1730	What if review

Day 5

0800	Science briefing
1000	Dive 8 (optional) equipment and situation dependant

Once you have completed your training, you are qualified to be a mission crewmember and you can suit up for your dive to the habitat. After entering through the moon pool, you will have the opportunity to store you scuba gear in the wet porch and take a warm shower before orienting yourself to the facility. One of the most important compartments is the Entry Lock, which contains communications

Figure 6.4. View from the interior of Aquarius. Aerospace engineer/aquanaut, Christopher Gerty, collects data for an experiment inside the Aquarius undersea habitat during the 13th NASA Extreme Environment Mission Operations (NEEMO) mission. Courtesy NOAA.

and life support equipment. You will also soon become familiar with the Main Lock, which is divided into a living and work area, and includes a bunk room containing six bunks. Other facilities include a galley, food storage areas, communications equipment, video equipment, and medical supplies. There are also plenty of viewports (Figure 6.4), so sightseeing is never a problem.

Inside the lab, aquanauts have access to a similar variety of systems that astronauts have onboard the ISS, such as data loggers and other data acquisition systems. Unlike NASA, however, which supplies its astronauts with company-issued laptops (IBM ThinkPads), aquanauts are required to bring their own personal computers. In common with astronauts working onboard the ISS, working onboard an inner-space station also presents some element of risk, although most accidents are preventable by using properly maintained equipment and judicious planning. Nevertheless, like any dangerous environment, there are some rules (Panel 6.4) that are in effect at all times.

Not surprisingly, a major activity while stationed onboard Aquarius is diving (Figure 6.5), which, just like a spacewalk, is planned meticulously. Every day, the dives must be reported to the Watch Desk using a Dive Plan Work Sheet, which

Panel 6.4. Non-negotiable rules

- No deviations from procedures outlined in aquanaut training and *Aquarius Operations Manual*
- No medications to be taken by aquanauts without prior approval of Aquarius Diving Medical Officer
- No smoking or open flames permitted in Aquarius
- Mission Control must be informed whenever an aquanaut leaves Aquarius, regardless of purpose or expected duration
- All equipment malfunctions to be reported to Watch Desk immediately
- All medical problems to be reported to Watch Desk immediately
- No alcoholic beverages or "social" drugs permitted

Figure 6.5. Most of every working day onboard Aquarius is spent diving, with aquanauts typically logging a 3-hr dive in the morning and another dive of similar duration after lunch. Courtesy NOAA.

details the names of the aquanaut teams, planned times out, destination, depth, route, expected times of return, air requirements, and any additional scientific materials required at work sites. For dives, bottom time is measured as time of departure from the hatch depth to return to the hatch. A typical dive day usually includes a 3–6-hr dive in the morning followed by a 4-hr interval in Aquarius, followed by a second dive in the afternoon of up to 3 hr.

At the end of a mission, decompression takes place in Aquarius and consists of slowly reducing the habitat's interior over a period of 17 hr. Once surface pressure is reached, aquanauts move to the Entry Lock and are recompressed to ambient water pressure before exiting via the wet porch. Following decompression, aquanauts are transferred to shore to undergo a 12-hr post-saturation observation period during which they must refrain from stressful activities and taking hot showers.

In addition to serving as an invaluable training tool for astronauts and helping scientists advance their research, Aquarius has also captured the imagination of the public. No doubt, the habitat will continue to revolutionize the way scientists work in the ocean. In doing so, it will continue to reshape the way we think about living there, but because it is a science facility, you must be a scientist (or astronaut) to visit, which means for those wishing to spend any length of time underwater, the only option is the Jules Verne Lodge. Fortunately, there may be more options available soon.

FUTURE RESORTS AND HABITATS

While staying onboard the Jules Verne Lodge or Aquarius provides an opportunity for undersea enthusiasts to gain their aquanaut certification, these habitats are hardly a step towards a more permanent human presence. Despite several visions of colonizing the ocean, no one has managed to realize the dream, but that has not stopped people from trying. In this section, we peer into the near future and take a look at the types of resorts you may be able to visit, assuming you have a healthy bank account!

Poseidon

If spending time in a scientific habitat is not your thing and you have a spare $30,000 lying around, you will soon have the chance to spend a few days in one of 24 luxurious suites 15 m underwater. Designed by US Submarine Engineering LLC and billed as the world's first underwater resort, Poseidon Undersea Resort (Figure 6.6) is currently under construction near Fiji and is expected to be open for business shortly after this book is published. After you have checked in, you will have the opportunity to learn to pilot a mini submarine, rack up some dives in pristine water, or simply watch the tropical fish. In common with any surface-based luxury resort, Poseidon will feature restaurants, bars, and retail boutiques in addition to a library, theater, and conference center. For those interested in tying the knot, there is even a wedding chapel!

This future vision of undersea opulence is the creation of Bruce Jones, president of US Submarines, a company better known for building custom tourist submarines.

Figure 6.6. Poseidon Undersea Resort is currently being constructed near Fiji. Courtesy US Submarines.

Jones designed Poseidon to provide guests with an all-inclusive vacation package that includes fine dining, stunning coral views, and opportunities to dive directly from the resort's airlock. Pressurized to 1 atmosphere, each undersea suite features large viewports built from polymethyl methylacrylate, a material usually used in submersible construction.

To save money, Jones is building the entire structure in a Portland shipyard before transporting it by barge to Fiji. Once there, the hotel will float in position while pilings are driven into the seafloor. Once the foundation is ready, the whole structure will be ballasted and sunk to the ocean floor. In common with so many underwater projects, Poseidon has had to overcome myriad logistical hurdles, such as funding and technical difficulties, and Jones still has a way to go before Poseidon opens its doors. There is even the possibility he will be beaten by a rival company in Dubai, where commercial developers are creating an even more ostentatious resort called Hydropolis.

Hydropolis

While it may look like the set of a science-fiction movie, Hydropolis (Figure 6.7) is about to become functional reality and, when operational, will incorporate innovations that will make it a truly unique underwater complex.

Figure 6.7. Hydropolis is the vision of developer and designer, Joachim Hauser. Due to open in 2010, suites will go for $5,500 a night. Courtesy Joachim Hauser.

Envisioned by developer and designer Joachim Hauser, the 220-suite resort can best be described as a geometric figure eight lying on its side inscribed in a circle. The spaces created in the basin will contain facilities such as restaurants, bars, and meeting rooms. To enter, visitors will begin at a land station, where they will board a train propelled by a fully automated cable along a modular, self-supporting steel guide-way. The upper level of the land station will house everything from a cosmetic surgery clinic to a marine biological research laboratory, while the lower level will be dedicated to staff rooms and goods storage. In common with Poseidon, Hydropolis promises to be a conceptual as well as a physical landmark – a vision Hauser hopes to achieve by incorporating many different elements associated with the ocean. By doing this, Hauser hopes to draw people's attention to the threats to marine life and inspire people to develop a new awareness of the ocean.

Dreams, however visionary, remain unfulfilled without the cash to support them and when the financial commitment exceeds $300 million (estimates suggest the final cost will be $490 million), those dreams become even more difficult to realize. After more than 2 years' work, Hauser received approval for the project by the Dubai Development & Investment Authority (DDIA), which established a framework to handle grants and authorizations. Once the financial backing was guaranteed, the sonar analysis of the seabed was completed, 7,000 anchors were sunk in place, and construction began. That was in 2005. The hotel was due to open in 2009, but various funding and technical difficulties have delayed its opening until 2010. When the 10-star-rated underwater hotel finally does open its doors, it will cost guests $5,500 a night. Incidentally, the cost of a night's stay includes protection from terrorists by the hotel's missile protection system!

Atlantica

Poseidon and Hydropolis will no doubt fill a niche in the adventure travel market, but the resorts still will not bring us closer to a permanent manned presence underwater. Fortunately, a group of aquanauts are working to remedy this by planning to embark upon an underwater expedition. Dubbed Atlantica, the mission will, if successful, set a new record for uninterrupted stay beneath the surface and set the stage for eventual permanent habitation. If all goes to plan, Dennis Chamberland, Claudia Chamberland, and Terrence Tysall will submerge in the Leviathan Habitat on August 2nd, 2010. Following them, rotating in 5-day visits, will be 24 other aquanauts, including scientists, teachers, and journalists. During their time underwater, the aquanauts will test systems and procedures for implementation in the much larger Challenger Station habitat scheduled to be launched in 2013 as the first permanent undersea colony off the Florida coast. Challenger Station, the largest manned undersea habitat ever built, will be populated with the first humans with no intention of calling dry land home again! It is possible they could become the first generation of people who live out their lives beneath the ocean.

The Leviathan Habitat has been designed for continuous habitation by four aquanauts during an extended undersea mission. Designed to maximize functionality in a restricted space, its floor plan includes two private staterooms, a private bath, a separate wet room with hot shower, and a command and control room. The habitat's life support system (LSS) is triple redundant, which means if any system should fail, there will still be two backups. In addition to being a fully functional scientific laboratory, the habitat's communication system features state-of-the-art digital communications, including a cell telephone service, high-speed internet, and even a satellite telephone. Following Leviathan's deployment will be Challenger Station, which will be home to a full-time complement of eight aquanauts – a number that will grow as modules are attached. Eventually, if Chamberland's dream is realized, Challenger will host the first human undersea colony.

Leviathan and Challenger are the result of the work of visionary, Dennis Chamberland, a man who has logged more than 30 days as an aquanaut and has designed six undersea habitats already! The former US naval officer has also worked as a nuclear engineer, and as a designer for advanced LSSs destined for the Moon and Mars. Claudia, Chamberland's wife, served as a platoon leader in the US Airborne before becoming an aquanaut in 1993, logging more than 30 days on more than a dozen undersea missions. Helping the Chamberlands during the Atlantica expedition will be Terrence Tysall, founder and president of Benthic Technologies, Inc., a consulting firm specializing in underwater applications and technology. Tysall also created the Cambrian Foundation, a federally recognized corporation dedicated to protecting aquatic resources. Tysall is also an avid diver, having gained instructor certifications with several diving organizations, and he serves on a variety of boards, including the Technical Advisory Board for the NOAA and the National Association of Underwater Instructors (NAUI).

In preparation for the undersea human colony, Chamberland founded the League of the New Worlds in 1990. The idea behind the League is simply to use the oceans as an analog for space exploration by forging the elements of the frontiers of the underwater world with those of the space environment. For members of the League, leaving the planet has a double meaning. One is living in space and the other is living underwater. It is a bold vision.

OCEAN OUTPOST

Undersea explorers often compare what they do with space exploration, although this comparison is rarely favorable to the astronauts, whom the aquanauts believe get far too much money. The disparity in funding between outer space and inner space is often mentioned in books on undersea exploration, which invariably include a comment about how humans know so much more about the surface of the Moon than the bottom of the ocean. Another favorite comparison is noting how more people have been into space (500 and counting) or walked on the Moon (12 and not counting!) than have reached the bottom of the sea (just two, and that was way back in 1960!). To be fair to the astronauts, these comparisons are a little misleading, since, every year, thousands of sailors travel under the oceans in submarines, thousands of commercial divers work on pipelines and oil rigs, and hundreds of researchers dive to great depths in the name of science. However, at the end of the day, the US[2] spends billions of dollars exploring space and far less money exploring the ocean. Fortunately, this spending disparity has not stopped those intent on colonizing the ocean, but how will they do it? Well, for people to permanently live underwater, one of the primary requirements is the development of a reliable, robust, and automated LSS and, in an environment as hostile as the ocean, such a task will not be easy. Since the greatest challenge to living for extended periods on the ocean floor is a dependable LSS, much of the remainder of this section addresses these requirements.

Life support system

The remote location of a habitat tens or hundreds of meters below the surface combined with the challenging aquatic environment presents unique challenges in terms of designing life support methods. Compounding the task is the limited

---

[2]   The NOAA Ocean Exploration and Undersea Research Program 2009 budget was $33.5 million (proposed). NASA's 2009 Budget was $17.9 billion, while the Pentagon's 2009 space budget was more than $25 billion!

Figure 6.8. The International Space Station has provided invaluable experience to engineers designing life support systems for future underwater habitats. Courtesy NASA.

experience in the field of extended-duration LSSs and subsystems, which, to date, is restricted to submarines, the ISS (Figure 6.8), the Space Shuttle, and Earth-based biospheres.

LSSs for an underwater habitat are very similar to those found in other human environments such as the ISS. This is not too surprising, since aquanauts need to breathe, eat, drink, and keep warm just like astronauts do. However, as engineers learned in the space industry, harsh environments have a habit of imposing great demands upon LSSs – a fact that has often resulted in astronauts spending an inordinate amount of time maintaining and repairing the equipment. It is because of the harshness of the undersea environment that habitats built to date barely warrant the designation and should instead have been classed as survival shelters due to the problems encountered in maintaining the integrity of the LLSs. Since aquanauts intending to spend hundreds of days underwater will want to focus on aspects of the mission other than repairing recalcitrant LSSs, it is important to ensure the system responsible for keeping the aquanauts alive is designed with adequate redundancy and built-in failure modes. But, before we consider the features of the underwater LSS, it is worthwhile considering the myriad functions an LSS must support (Panel 6.5).

Panel 6.5. Life support systems functions

Maintain environment
Control atmosphere total pressure: prevent over-pressurization and/or under-pressurization, equalize atmosphere pressure, control metabolically inert gas partial pressure, and add metabolically inert gas to atmosphere
Control oxygen partial pressure: add oxygen to atmosphere
Control atmospheric temperature: remove or add sensible heat
Control atmospheric humidity: remove or add moisture
Control uniformity of atmospheric composition: ventilation velocities in the habitat volume and exchange atmosphere between modules
Control partial pressures of atmospheric contaminants: remove gaseous atmospheric contaminants
Control airborne particulates: remove airborne particulates
Control microbes: remove airborne and surface microbes

Respond to emergencies
Respond to rapid decompression: detect and recover from rapid decompression
Respond to fire: detect, isolate, suppress, and recover from fire
Respond to hazardous atmosphere: detect and recover from hazardous atmosphere

Provide resources
Provide inert gas: supply/store inert gas and accept external inert gas
Provide oxygen: supply, store, regenerate oxygen, and accept external oxygen
Provide water: supply, store, regenerate water, and accept external water
Provide food: store and process food and food ingredients
Manage gaseous wastes: accept, transport, store, process gaseous wastes, and dispose of excess gaseous wastes
Manage wastewater: accept, transport, store, process wastewater, and dispose of excess wastewater
Manage solid and concentrated liquid wastes: accept, transport, store, process solid, and dispose of solid and concentrated liquid wastes

Maintain thermal conditioning
Accept thermal energy
Transport thermal energy: transport excess thermal energy to cooling external interface
Release thermal energy: reject excess thermal energy and reuse thermal energy
Cool equipment and maintain surface touch temperatures:
    Transport equipment thermal energy loads
    Prevent water condensation

From Panel 6.5, it is easy to see that the habitat's LSS must take into account significant constraints that accommodate human variability. To ensure the crew's health and working capacity in a sealed environment, conditions must be similar to the conditions to which the crew is accustomed. A slight deviation of the environment from permissible limits adds stress and taxes physiological systems that, in turn, may compromise work performance. To ensure an optimum environment is maintained, the LSS includes several systems and subsystems, the interaction of which is described in Table 6.3.

Table 6.3. Life support subsystem description.

HABITAT LIFE SUPPORT SUBSYSTEMS AND INTERFACES

Subsystem	Description	Life support interfaces
Air	Stores and maintains the habitat atmospheric gases, pressure, and composition, and serves as a fire detection and suppression system	Biomass, food, thermal, waste, water, dive support, human accommodations, integrated control, power
Biomass	Produces, stores, and provides raw agricultural products to the food subsystem while regenerating air and water	Air, food, thermal control, waste, water, integrated control, power
Food	Receives harvested agricultural products from the biomass subsystem, stabilizes them, and stores raw and stabilized agricultural products, food ingredients, and prepackaged food and beverage items. Transforms raw agricultural products into a ready-to-eat form via food processing and meal preparation operations	Air, biomass, thermal, waste, water, dive support, human accommodations, integrated control, power
Thermal	Maintains cabin temperature and humidity within bounds and rejects collected waste heat to environment	Air, biomass, food, waste, water, human accommodations, integrated control, power
Waste	Collects and conditions solid waste material from anywhere in the habitat, including packaging, human wastes, and inedible biomass	Air, biomass, food, thermal, water, dive support, human accommodations, integrated control, power

Subsystem	Description	Life support interfaces
Water	Collects wastewater from all sources, recovers and transports potable water, and stores and provides that water at purity for crew consumption and hygiene	Air, biomass, food, thermal, dive support, human accommodations, integrated control, power

EXTERNAL LIFE SUPPORT INTERFACES		
External interface	Description	Life support interface
Diving support	Provides life support consumables for diving, including oxygen, water, food, and carbon dioxide and waste removal	Air, food, waste, water, human accommodations, integrated control, power
Human accommo-dations	Includes crew cabin layout, crew clothing, laundering, and crew interaction with life support system	Air, biomass, food, thermal, waste, water, diving support, integrated control, power
Integrated control	Provides appropriate control for all the life support system	All
Power	Provides energy to support all equipment and functions within life support system	All

As you can see, designing an underwater LSS is not easy. Making the task even more difficult is the requirement that the LSS must work reliably underwater for hundreds of days. One of the mistakes of many mission designs is the assumption that high-performance LSSs will function flawlessly for the duration of the mission. However, experience onboard the ISS has shown that despite extensive (and expensive!) ground checkout, hardware failures still occur with frustrating regularity, so for a long underwater mission, it will be necessary to include a backup cache of replacement units.

So, we have looked at the LSS functions. Now, before we examine the various components of the LSS, we need to look at the requirements.

Before sitting down and designing a long-duration LSS, it is necessary to present best estimates of what may be required to support the crew during their stay on the ocean floor. A detailed discussion of all the LSS elements is beyond the scope of this book, so what follows is an estimate of some of the primary LSS elements. Let's take a look at breathing gas first.

To date, the breathing gas of underwater habitats has been air, nitrogen–oxygen, or helium–oxygen. Of these, air is the cheapest and most convenient, and is safe to a depth of 15.2 m for SAT exposures up to 1 week (breathing air longer at this depth may cause respiratory problems), while for depths between 15.2 and 36.6 m, the most suitable gas is a normoxic mix of nitrogen and oxygen. "Normoxic" is a physiological term describing a breathing gas mixture that contains 21% oxygen. A normoxic mix (Panel 6.6) is crucial in the design of an underwater LSS when decompression and nitrogen narcosis are considered because saturation at a given depth using a normoxic mix is equivalent to a greater depth than if air were used (Table 6.4).

Table 6.4. Equivalent depths for air and normoxic breathing gases.

Air depth (m)	Air PO_2 ATA	Equivalent normoxic depth (PO_2 = 0.2095)
0	0.79	0
10.7	1.85	8.4
22.9	3.06	18.0
32.3	3.97	25.3
42.7	5.03	33.7

Source: Miller, J. (ed.), *NOAA Diving Manual: Diving for Science and Technology*, 2nd edn. Government Printing Office, Washington, DC (1979).

Panel 6.6. Normoxic gas mix

Oxygen must be present in every breathing gas, whether to support underwater habitats or spacewalking astronauts. This is because it is essential to the human body's metabolism and the body cannot store oxygen for later use. In fact, unless you happen to be a freediver, if your body is deprived of oxygen for more than a few minutes, unconsciousness and death are guaranteed! In underwater use, the oxygen fraction of a breathing gas mixture is often used when defining the mix. For example, *hypoxic* mixes such as trimix and heliox, which are used in technical diving, contain less than 21% oxygen (often a boundary of 16% is used), while *normoxic* mixes contain 21% oxygen – the same fraction as found in air.

One drawback to using a normoxic mix is that the maximum operating depth (MOD) could be as shallow as 45 m. This is because the fraction and pressure of the oxygen determine the deepest that a gas mix can safely be used to avoid *oxygen toxicity*. The fraction and pressure are expressed by the partial pressure of oxygen (ppO_2) and the minimum safe ppO_2 in a breathing gas is held to be 16 kPa (0.16 bar). Below this ppO_2, the diver may risk unconsciousness and death due to hypoxia (depending on individual physiology and exertion level), while the maximum safe ppO_2 in a breathing gas is typically between 100 kPa (1 bar) and 160 kPa (1.6 bar), although for dives of less than 3 hr, the threshold is 140 kPa (1.4 bar). Above this level, the diver risks oxygen toxicity, including a seizure.

Another problem with using air as a breathing gas is nitrogen narcosis, which manifests itself in most divers at depths exceeding 45 m, equivalent to 36.6 m when breathing a normoxic mix of nitrogen and oxygen (Table 6.4). Because of this, most SAT diving below 45 m has used helium as an inert gas in place of nitrogen. However, while helium has no narcotic effect, its high thermal conductivity means body heat loss is significant, which means aquanauts would need their habitats heated as high as 32°C! Another disadvantage of helium is the communication problem it causes, which results in divers sounding as if they are trying to imitate Donald Duck! Fortunately, thanks to the latest electronic helium unscramblers, garbled helium speech can be made intelligible.

Once the breathing gas has been chosen, the next consideration is how to scrub it of carbon dioxide (CO_2). Highly soluble in body tissues, CO_2 is the end point of metabolism of oxygen. If a diver accumulates an excessive amount of CO_2, his/her performance may be affected detrimentally. For example, since CO_2 is a narcotic gas, it is capable of depressing awareness, including loss of consciousness, even at relatively low concentrations. At severe elevations, the gas may cause surgical anesthesia! Given the severe consequences of elevated CO_2 levels, it is not surprising that more than two dozen methods have been suggested for removing the gas from closed environments. Until recently, one of the most widely used means of removing CO_2 from underwater habitats was to use absorbents such as soda–lime. While soda–lime is reasonably effective, its performance is influenced by factors such as temperature and humidity and if environmental conditions are not maintained within narrow margins, the absorbent may cause water to condense in the absorbent, which renders the absorbent inactive.

Another way of removing CO_2 is to use solid lithium hydroxide (LiOH), which is used in space vehicles. On board the Space Shuttle, for example, CO_2 is removed from the air as it passes through a pair of LiOH canisters. Since LiOH is an efficient CO_2 absorbent, it reacts with gaseous CO_2 to form solid lithium carbonate and liquid water. One disadvantage of using LiOH is that is not regenerable, which means aquanauts would have to pack a number of canisters with them on their way to the habitat. These canisters would then have to be replaced every so often, which would require several resupply missions to the habitat.

A more effective way is to use a two or four-bed molecular sieve. It is a system that is used on the ISS and on board naval submarines. The two-bed molecular sieve (2BMS) removes CO_2 from a wet gas stream by passing the gas stream through an integrated absorption bed. The absorption bed comprises the first portion, which is an absorbent for water whereby water is removed from the gas stream, a second portion, which is an absorbent for CO_2 whereby CO_2 is removed from the gas stream, and a third portion, which is an absorbent for water that is saturated with water whereby water is reintroduced into the gas stream. The process may be operated continuously by employing two integrated absorption beds, one of which is absorbing CO_2 while the other is desorbing CO_2.

In addition to removing CO_2, the LSS must also monitor contaminants. This is the function of the Trace Contaminant Control Subassembly (TCCS), which maintains atmospheric concentrations of trace contaminants within acceptably safe levels. Three main components comprise the TCCS: an expendable activated carbon bed, a thermal catalytic oxidizer, and an expendable post-sorbent bed. Additionally, the TCCS contains a blower, flow meter, and electrical interface assembly. The process flow rate is forced through the carbon bed, which contains granular-activated carbon treated with 10% phosphoric acid that prevents ammonia from entering the thermal catalytic oxidizer. Upon exiting the carbon bed, the process stream splits. Approximately one-third of the air flows through the thermal catalytic oxidizer and post-sorbent bed before rejoining the bypass stream just before the system exhaust. Principle parts of the thermal catalytic oxidizer include a recuperative heat exchanger, an electric heater, and a catalyst bed. The catalyst bed contains a platinum-group metal catalyst supported on alumina pellets. After passing through the catalytic oxidizer, the process stream passes through a bed of lithium hydroxide to remove any acidic oxidation products.

The characteristics of the underwater habitat environment will make the task of regulating environment temperature and humidity a challenging one, since the Thermal and Humidity Control System (THCS) must account for significant heat generated by human activity and the habitat's electrical equipment. The habitat's thermal control may be achieved through passive and active mechanisms. Passive mechanisms promote thermal stability using non-moving components, while active thermal control systems (ATCS) mechanically transport heat from the habitat's interior to its surface. Once there, heat is radiated into the ocean or rejected using evaporative cooling. ACTS mechanisms usually comprise heat exchangers and heat-transfer loops. Heat exchangers transfer heat from one medium to another, while heat transport loops contain a heat-transfer medium, such as a fluid or gas. Pumping the heat-transfer fluid through the heat-transfer loop mechanically transports the contained heat to the desired location. From there, a heat exchanger can be used to transfer heat to a secondary heat-transfer loop or to reject the heat.

In addition to the heat problem, there is also a water problem due to metabolic activity that adds significant amounts of water to the atmosphere. Through perspiration and respiration, humans contribute approximately 2.3 kg of water vapor per day to the atmosphere. It's a lot of moisture, and it has to be removed to maintain a comfortable environment and to minimize moisture condensation and adsorption onto materials in the outpost environment. Excess water vapor can be removed in several ways. It can be physically adsorbed, chemically adsorbed, or condensed onto cold surfaces. For an underwater outpost, atmospheric humidity removal is probably best achieved using regenerative systems that use molecular sieves as desiccant beds to desiccate the air before passing it through CO_2-scrubbing sieves.

Ventilation also plays a critical role in the maintenance of a uniform outpost temperature. A ventilation system, which comprises a combination of fans, air ducts, and intake and diffuser grates, is required to facilitate transfer of atmospheric heat and humidity to the heat exchanger, which forms part of the ACTS.

Another important LSS feature will be the food subsystem. Given the remote location and demanding environmental conditions faced by crewmembers several tens of meters underwater, food will play an important psychological role and providing palatable and enjoyable food will be important in maintaining crew morale. The food supply subsystem will provide crewmembers with the proper amount of daily protein, carbohydrates, and fat during their mission. To help designers, it is likely the following assumptions will be made:

1. Food will be allotted by daily allowance. Trading one day's meals for another will be permitted, but consumption will be limited to the equivalent of one "daily allowance" per day
2. Each daily allowance will meet minimum USRDA levels of vitamins and minerals
3. A 10% buffer for each person/day will be added to the baseline daily requirements for protein, carbohydrates, and fat. This assumption will provide:
   a. Additional food for variance in kcal/person intake
   b. Additional food for contingency
   c. Additional food for variance in eating habits between crewmembers
4. Packaging mass assumed to be 0.5 kg/day of food
5. Water to prepare or re-hydrate any food will be provided by the water management subsystem

In common with ISS operations, food will probably be categorized as either *menu food* or *pantry food*. Menu food will consist of three meals per crewmember per day, while pantry food will comprise a 2-day contingency food supply that will also contain food for snacks and beverages between meals and for individual menu changes. Types of food will include fresh, thermostabilized, rehydratable, irradiated, intermediate-moisture, and natural-form food and beverages. Foods packaged in rehydratable containers include soups, casseroles like macaroni and cheese, appetizers such as shrimp cocktail, and breakfast items like scrambled eggs. In contrast, thermostabilized foods are prepackaged for one serving and include products such as grilled chicken and ham, tomatoes and eggplants, or puddings. Intermediate moisture food, on the other hand, describes food items that are preserved by restricting the water available for microbial growth while retaining sufficient water to give the food a soft texture and allow it to be eaten without further preparation. Food items representing this category are dried peaches, pears and apricots, and dried beef.

Obviously, the food subsystem will be located in the galley, which will be a multipurpose facility providing a centralized location for crewmembers to prepare meals. In addition to having facilities for heating food, rehydrating food, and stowing food system accessories and food trays, the galley will also feature a rehydration station, oven/food warmer, food trays, and food system accessories.

Inevitably, after discussing food, the issue of waste management needs to be addressed. An underwater waste management system (WMS) will operate on the principles of reducing waste volume, waste safening and stabilization, and odor control. Reducing waste volume is simply accomplished via mechanical compaction methods, while safening and stabilization can be achieved through processes such as containment and drying. The job of controlling waste odors, on the other hand, is performed by the TCCS.

With all these systems and subsystems functioning, it becomes increasingly difficult to keep noise levels to an acceptable level, but this is an important requirement, since activities such as communications, sleeping and rest, and mental concentration are adversely affected by noise. In common with ISS operations, outpost rules will probably call for hearing protection when noise levels reach 74 dB. This is important, since noise generation will need to be controlled to reduce chance of personal injury, communication interference, fatigue, or ineffectiveness of the man–machine interface.

Diving procedure

One of the attractions of living underwater will obviously be diving, but diving from an ocean outpost will demand slightly different diving practices from those normally employed by regular scuba-divers. For example, essential items of equipment for a scuba-diver include the buoyancy compensator device (BCD) and a weight belt, but for a saturated aquanaut, safe diving practice would be to *not* wear these, since one of the most serious dangers is surfacing accidentally. In fact, given the risks of diving from a SAT outpost, it is likely dives will be limited to lateral excursions only. Also, due to the aquanaut's reliance upon the outpost as the only source of life support, excursions will need to be planned meticulously to avoid the risk of becoming lost. This will be achieved by laying out a grid of lines, markers, and direction indicators around the outpost to provide divers with plenty of location reference points. Of course, even with the best prior planning and preparation, emergencies are bound to happen. A diver might get lost, run out air, or suffer an equipment malfunction. To assist a diver in distress, the outpost and its perimeter will feature emergency breathing stations, strobe lights, and sonar locators placed at regular intervals. These features will be particularly important at the start of outpost operations, since the aquanauts will be very busy (Table 6.5) with construction tasks and the terrain will be unfamiliar. Gradually, as the outpost becomes established, the outpost's inhabitants will focus more on exploration and the search for resources, although these activities will still require sustained periods of activity outside the protective environment of the outpost.

Dives from the outpost will be classed as either scheduled, unscheduled, or contingency. *Scheduled* (planned) dives will be conducted to accomplish tasks to support specific mission operations, while *unscheduled* (unplanned) dives will be conducted to accomplish tasks not included in the mission timeline but will be needed to achieve mission success or to repair/override failed systems. The final class, the *contingency* dive, will be conducted to ensure the safety of the outpost. The subject of contingency dives leads us to the subject of safety – an issue that will be key to the success of the outpost. The outpost's safety will lie in the utility of the outpost as a safe and efficient system for supporting humans under the ocean and one of the most critical aspects of ensuring that safety is sustained will lie in the aquanaut's ability to cope with and recover from emergency situations. If one or more emergency conditions should arise, aquanauts must be capable of quickly initiating and successfully executing procedures to mitigate the failures and recover. In all likelihood, these procedures will need to be performed under tight timing constraints. To help develop contingency procedures, aquanauts will no doubt perform emergency drills inside an outpost simulator, using techniques used by astronauts preparing for ISS increments. Using videotapes of aquanauts performing emergency procedures in the outpost will provide human factors engineers with valuable information on how emergency procedures can be best executed and provide insight into what kind of aquanaut–procedure interaction is more prone to error. This information will be fed back into the procedure correctness criteria to make them more comprehensive and effective.

Table 6.5. Diving tasks during outpost missions.

Task	Description	Task	Description
Site preparation	• Survey and stake-out • Smoothing • Establish navigation aids • Transport modules to site	Power and thermal control systems	• Site preparation • Unload equipment • Deploy and assemble • Connect to distribution
Logistics	• Unload and unpack • Transport • Transfer • Storage	Science	• Sample collection • Installation of experiments • Location of mapping for geological survey
Resource operations	• Resource process site set-up (pressure vessel, plumbing, gas holding tanks, pumps, heat exchangers)	Upkeep	• Inspection • Field checks/measurements • Replacement/repair of systems/subsystems

One emergency procedure that will surely be practiced on a regular basis will be the response to fire. A fire in the enclosed environment of an underwater outpost would be disastrous. Although the likelihood of a major fire will be reduced by minimizing atmospheric oxygen content and by using fire-retardant materials, crewmembers will still need to have a high degree of familiarity with the outpost's fire suppression systems.

Other systems

Just as the threat of emergencies is a fact of life when living in any extreme environment, so too is the possibility that one or more crewmembers may suffer minor illnesses and/or injuries from time to time. Due to the decompression penalty, it will not always be possible for severely injured or ill crewmembers to be returned to the surface, which is why the medical subsystem (MSS) will be one of the most important components of the outpost's LSS. The MSS will likely consist of two separate packages: the medications and bandage kit (MBK) and the emergency medical kit (EMK). The MBK will contain oral medications consisting of pills and capsules, bandage materials, and medications, while the EMK will contain medications to be administered by injection, items for performing minor surgeries, diagnostic/therapeutic items, and a microbiological test kit for testing for bacterial infections. Another important feature of the MSS will be a telemedicine capability, which will be crucial in the event of a serious injury requiring surgical intervention. By this method of surgery, a surgeon in a remote location controls the robotic instruments performing the actual surgery. It is a method that has been practiced successfully over intercontinental distances, and has also been investigated by NASA astronauts conducting research aboard Aquarius.

AQUANAUT SELECTION

"Men wanted for hazardous journey. Low wages, bitter cold, long hours of complete darkness. Safe return doubtful. Honour and recognition in event of success."

Advertisement rumored to have been posted by Sir Ernest Shackleton
before the launch of his legendary 1914 Imperial
Trans-Antarctic Expedition

The advertisement placed by the great Shackleton may be apocryphal but its content applies equally to those selected for underwater missions – an adventure characterized by a host of physiological and environmental stressors. The expeditions embarked upon by explorers such as Shackleton almost a century ago resemble in many ways the conditions of isolation and confinement that will be experienced by future aquanauts residing on the ocean floor. The conditions will be different, but many of the problems confronting future aquanauts will be the same ones that troubled explorers in the past – a reality that will be reflected in the unique selection criteria applied to those chosen to live in ocean outposts. This section describes how aquanauts may be selected and trained, how undersea agencies will decide who has the "right stuff" for an outpost mission, and what factors, beyond technical skills and education, should be considered for selecting a crew for what will be the most arduous undersea human mission to date.

In common with the crewmembers of Shackleton's expedition, the austere and isolated conditions facing future aquanauts will impose significant hardship upon those selected. While it will be assumed an aquanaut has the skills and knowledge necessary to perform the duties of a crewmember, these abilities will count for nothing if he or she cannot get along with others for several months in the confines of a habitat that may be no larger than an apartment! Given the unique characteristics of such a mission, the issues of crew composition and crew compatibility clearly become factors almost as important as the selection process itself. Due to the potential for these issues to impact negatively upon a mission, it is important to address them before moving on to discussing the selection process itself.

It is likely the size of the first crew will fit with the current belief that smaller is better. Such a policy was implemented on many of the most successful polar expeditions such as Shackleton's Imperial Antarctic Expedition, which consisted of just 27 members. Despite extreme isolation and prolonged confinement, Shackleton's expedition was characterized by few interpersonal problems, thanks largely to the small homogeneous crew – a lesson unlikely to be overlooked when it comes to defining the composition of a long-duration undersea crew.

The occupational role of each undersea crewmember has yet to be determined, but it is likely, given the extended duration of the mission, one crewmember will be a medical doctor. The role of commander will probably be assigned to the crewmember with the most underwater experience and, given the science objectives of such a mission, it is inevitable the crew will include at least one scientist, while other crewmembers will be cross-trained in various scientific disciplines.

The issue of whether a crew should be all-male, all-female or mixed remains a contentious matter. Some have argued a female crew would exhibit preferable interpersonal dynamics and be more likely to choose non-confrontational approaches to solve interpersonal problems, whereas others have made a case for a mixed crew, claiming crews with women are characterized by less competition and seem to get along better. Evidence from Antarctic winter-over crews supports each of these arguments and suggests the inclusion of women in underwater crews would serve a socializing purpose, in addition to their mission function. However, the introduction of a single female into a male group may have destabilizing effects because of sex issues. For example, what effect would a passionate affair during a long-duration underwater mission have upon other crewmembers and on overall crew performance?

The next issue is crew compatibility, which has often been viewed as an opaque process because there is no one measure to predict whether a crew will work together effectively. Some researchers favor the use of psychological performance tests and personality questionnaires, whereas other investigators prefer a more behavior-oriented approach. The Russians, on the other hand, who have invested considerably in developing methods to assess interpersonal compatibility, consider biorhythms to be a useful tool for selecting cosmonauts. Once again, useful lessons can be found in the annals of polar exploration and research conducted during Antarctic winter-over increments that suggest personality traits and interpersonal skills be carefully scrutinized when selecting crews for long-duration missions. However, all the psychological tests in the world will be unable to predict how crewmembers will interact. To resolve this, a candidate crew may spend time in a high-fidelity outpost simulator to demonstrate to themselves and mission managers they can adapt to the many unique stressors associated with living and working in close proximity.

A veritable cornucopia of knowledge regarding crew composition and compatibility exists thanks to the successful expeditions of Shackleton, Nansen, and Amundsen and experience from Antarctic research stations. This history of crew dynamics in harsh environments provides mission planners with more than enough information to carefully select a compatible crew. But what does the actual selection process entail?

Crew selection

The agency with perhaps the greatest experience in crew selection is NASA, so it is likely selection procedure will follow NASA's current basic health and education selection requirements, with all crewmembers having either a science or engineering background. Crewmembers will also be required to meet a number of unique mission requirements, some of which are summarized in Table 6.6.

Table 6.6. Selection requirements.

Personal and medical requirements	
Male	Age: > 50
Meet underwater medical standards	Possess sense of community
Free of psychological problems	Possess effective conflict resolution skills
Technically competent	Possess sense of teamwork
Social skills and behavioral traits	
Social compatibility	Tolerance
Emotional control	Agreeable and flexible
Patience	Practical and hard-working
Introverted but socially adept	Does not become bored easily
Sensitive to the needs of others	Desire for optimistic friends
High tolerance for lack of achievement	High tolerance for little mental stimulation
Self-confident without being egotistical	Subordination of own interests to team goals
Crew compatibility traits	
Tactfulness in interpersonal relations	Effective conflict resolution skills
Sense of humor	Ability to be easily entertained

Whether Chamberland's vision of a permanent undersea colony is realized remains to be seen. On paper, underwater cities would certainly go a long way to dealing with the over-population concerns now taking shape across the world. While many in the space agencies and the corridors of power look to the stars to seek answers from soaring numbers in cities and towns, few take the time to look below the surface, which is perhaps where the answer truly lies. Over 70% of the world is covered by water, so making use of it would certainly make sense in proportion to the land mass available. However, even though we are at the beginning of the twenty-first century, underwater outposts still has the ring of science fiction. From the practical aspect, the massive amounts of spending required to build and maintain an underwater complex would mean that it would undoubtedly require the combined efforts of several countries. Also, unless there is a breakthrough in hydroponics, most of the food supplies would have to be imported via supply subs, as the lack of sunlight would result in little or no photosynthesis to grow foodstuffs. Another avenue for the underwater population concept may lie in the industrial sector. Fossil fuels aside, valuable resources such as diamonds and rare minerals are known to exist underwater, so if an area was found where there was an abundance of such resources, it would prove a strong case for development of underwater mining operations. The cost of running such an operation would likely be exceeded by the bounty reaped from the depths, so it is worthwhile examining just how much of value is down there.

Section III

Ocean Exploitation

The oceans have much to offer of economic importance, and as energy use continues to increase, the unexploited reserves held in the seabed have become a target for offshore oil companies seeking to capitalize on this extensive reservoir of marine resources. But, perhaps more important than the exploitation of energy reserves is the search for marine-derived medication. Marine plants and animals have biotechnological potential in the treatment of a wide variety of human illnesses. For example, coral reefs, often labeled as the rainforests of the sea, contain unique chemicals that can be used to fight cancer, diabetes, and other diseases. While chemicals from land-based plants and microbial fermentation are on the decline, scientists have barely scratched the surface of the ocean's medicinal potential.

7

Deep-Sea Mining and Energy Exploitation

In James Cameron's alien first-encounter movie, *The Abyss*, much of the action takes place on a semi-mobile drilling platform called The Benthic Explorer, located hundreds of meters beneath the surface. While present-day technology may not match Cameron's *Benthic Explorer*, mining the ocean floor is a step closer to reality thanks to increased energy demand and the efforts of neophyte marine mining companies who are exploring the possibility of mining deep-sea-floor deposits. Until very recently, such a venture was neither economically nor technically possible, but recent advances in technology mean we may be on the brink of an era of deep-ocean exploitation.

BLACK SMOKERS

The transformation in ocean exploitation is being spearheaded by Nautilus Minerals and Neptune Minerals. While Neptune is assessing deposits (known as Kermadec) to which it holds the rights in territorial waters off the north coast of New Zealand's North Island, Nautilus is collecting samples from a deposit (known as Solwara 1) to which it holds the rights in the Bismarck Sea off the coast of Papua New Guinea. Each company is banking on the economic potential of undersea deposits of polymetallic sulphides, sulphur-rich ore bodies produced in volcanic regions by "black smokers". Black smokers (Figure 7.1) are formed when seawater seeps into the porous ocean floor, is heated, and re-emerges through vents carrying dissolved minerals (Figure 7.2). When the hot water makes contact with the cold ocean water, minerals precipitate, creating black smokers that, over time, collapse and accumulate to form valuable ore deposits known as sea-floor massive sulphide (SMS) deposits. Over time, the hydro-thermal plumbing systems close off and migrate to new fracture systems on the sea floor. The challenge for companies like Nautilus and Neptune is to find the cold, inactive SMS deposits on the sea floor – a difficult task, since the deposits cannot be traced by following the smoke plumes, as happens in active SMS vent fields.

One of the first mining geologists to explore the economic potential of black

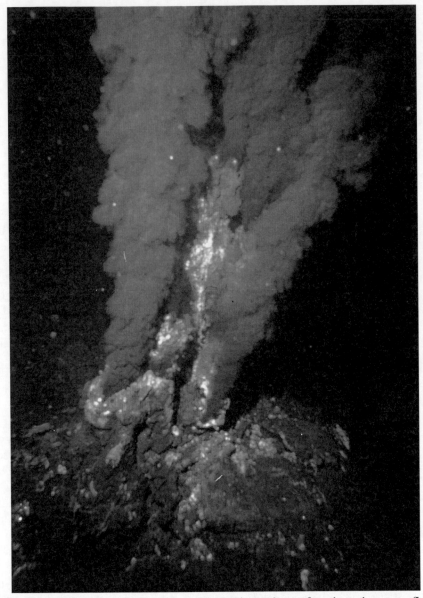

Figure 7.1. A black smoker is a type of hydrothermal vent found on the ocean floor. They are formed when superheated sulfide-rich water from below Earth's crust comes through the ocean floor. When the water comes into contact with the cold ocean water, minerals precipitate, forming a black chimney-like structure. Black smokers were discovered in 1977 by scientists from Scripps Institution of Oceanography using Alvin. Black smokers exist in the Atlantic and Pacific Oceans at an average depth of 2,100 m. The water at a vent may reach 400°C, but despite the temperature, the water does not boil because the water pressure at that depth exceeds the vapor pressure. Courtesy NOAA.

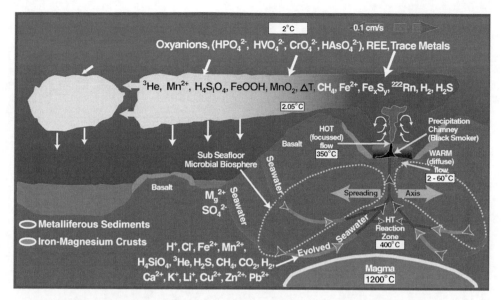

Figure 7.2. Diagram of the hydrothermal circulation in a mid-oceanic ridge (MOR) system. Courtesy NOAA.

smokers was Dr. Scott, who joined members of the Scripps Institute of Oceanography and the Woods Hole Oceanographic Institution (WHOI) in the Alvin submersible to explore newly discovered black smokers 2 km below the surface in the Gulf of California. Dr. Scott, who is the Director of the Scotiabank Marine Geology Research Laboratory, points out that after more than 20 years of promoting the possibility of mining the deposits created by black smokers, the launch of ventures by Nautilus and Neptune represents a move into new territory. Back in 1982, when Dr. Scott ventured into the Gulf of California, mining companies did not even entertain the possibility of mining at such depths, but pioneers such as Nautilus and Neptune realized that it is a lot easier to go down a couple of thousand meters of water than through a couple of thousand meters of rock. While the technology to reach such depths exists today (some oil wells are located at 2,500 m depth in the Gulf of Mexico), the challenge for undersea mining companies will be developing the technology to extract the ore. To do this, deep-sea mining machines and semi-submersible platforms are being developed, similar to the ones currently used in the oil industry. It sounds expensive, but Nautilus and Neptune figure the investment will be worthwhile. That is because the SMS deposits are rich in copper, zinc, and lead with a high gold and silver content. In their quest for the SMS deposits, both companies are searching in the so-called "ring of fire" – a region known for sea-floor earthquakes and volcanic eruptions. It may sound dangerous, but this sort of environment is perfect for creating SMS deposits, some of which can weigh as much as 1 million tonnes. The advantage of this type of mining is that instead of having to find one large individual deposit of ore body, as with land-based mines, a marine mining operation can be moved from deposit to deposit. The ventures promise to be lucrative. For example, Technip, a French engineering firm for the oil and gas industry, was contracted by Neptune to investigate the profitability of the company's operations. It found that Neptune's SMS deposits had a value of US$500–US$2,000 per tonne. When you consider that Neptune's operating costs per tonne are estimated at US$145–US$162, you begin to get an idea of the profit potential.

Of course, any enterprise dealing with the deep sea comes with an overhanging environmental question mark. But, if all the talk of robotic excavating machines has you wondering about the impact on the underwater environment, you needn't worry, since ocean-floor mining sidesteps many of the problems normally associated with its terrestrial cousin. First, there is no acid drainage because the acids are neutralized by the alkaline sea water and, second, because SMS deposits are on the ocean floor, there is no requirement for excavation, so no permanent structures are left behind. Finally, the mining does not touch active black smokers, so these regions of rich submarine life are preserved.

OIL

Another facet of future underwater exploitation will be oil. Exploration and production in Northern Europe are moving into ever deeper water, which has stimulated the development of subsea production systems with remote operation and maintenance that demands and will continue to demand manned intervention. In turn, this has required the development of advanced diving procedures and technology to permit safe and productive saturation (SAT) diving to depths as deep as 500 m. Unfortunately, diving to these depths is anything but safe, and ethical concerns have questioned the efficacy of the safeguards in place to protect divers from bone necrosis (Figure 7.3) and neurological disorders that are common occupational hazards among deep-diving SAT divers (Panel 7.1).

Panel 7.1. Long-term consequences of deep diving

Divers have dived as deep as 686 m of seawater (msw) and, as we have seen in Section I, these SAT divers regularly work at depths exceeding 200 msw. While the divers can work productively at these depths, research has shown that working as a deep-diving SAT diver is anything but healthy. Scientists studying SAT divers have found that the increased pressure causes all sorts of changes to the bone and in the central nervous system (CNS). Some divers experience hand tremors, postural instability, and gastrointestinal problems, while others may suffer from somnolence and cognitive dysfunction [1–3]. The symptoms, which are highly correlated with deep SAT diving and decompression sickness (DCS), have been termed the high pressure neurological syndrome (HPNS).

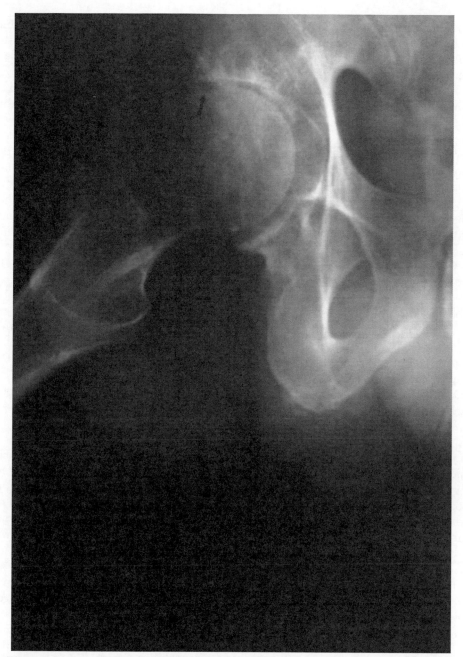

Figure 7.3. Lateral view of the right hip in a patient with avascular necrosis. Dysbaric osteonecrosis is a form of avascular necrosis suffered by saturation divers. The condition occurs as a result of the death of a portion of the bone that is thought to be caused by nitrogen embolization (blockage of the blood vessels by a bubble of nitrogen coming out of solution) in divers. Courtesy Wikimedia.

Despite the risks, SAT divers will continue to clock on for their well paid but hazardous job, just like they have done since the mid 1960s. This is because the offshore oil industry is booming once more, as evidenced by the increase in orders for diving support vessels built for offshore contractors who are moving into ever deeper waters and harsher environments. For example, SubSea 7, a UK-based subsea construction and engineering company, has the largest and most capable diving support vessel in the market, while the Dutch vessel, *The Seven Atlantic*, will be able to deploy two 24-man saturation diving systems designed to operate at 350-m depth in the North Sea. Meanwhile, France's *DSV Acergy Havila* will be one of the most capable vessels of its kind when it is delivered in 2010, since its 24-man saturation diving system will be rated to operate at 400 msw. These systems move SAT diving into the twenty-first century with computerized monitoring systems that not only eliminate many of the valves and gauges that once featured in such systems, but also provide continuous monitoring of the health of the diving team during decompression.

The offshore oil boom is just as strong on the other side of the Atlantic. The US Strategic Oil Reserve (SOR) is a government-controlled oil reserve which, in times of economic emergency, can be called upon to keep the flow of oil to the refineries. Located in four cavernous reservoirs in the Gulf of Mexico, the reserve holds more than 700 million barrels-worth of oil – a volume sufficient to supply the US with fuel for a little more than a month. It is little more than a stop-gap measure. However, if future oil supplies become erratic, one short-term solution could be the implementation of petrol rationing, which might be followed by calling on the SOR to alleviate the shortage. To avoid this scenario, the US is embarking upon aggressive oil exploration and since most of the accessible fields have already been drained dry, companies now have their sights set on the last frontier of oil and gas exploration – the Arctic and Antarctic.

The basins of the Arctic and Antarctica are where, according to many academics and geologists, the last 25% of the world's remaining resources can be found. Exploitation of resources in these locations, where the environments are even more hostile than the North Sea, will be a difficult and challenging prospect. Although no oil or gas prospecting has been carried out to date, survey results show great promise for the region and as the global price of oil rises, options such as the Antarctic will become more economically viable as the pay-off increases over the cost to extract. Now, you may be wondering how the Antarctic can be exploited for oil when the wilderness is protected by the Antarctic Treaty *and* the Madrid Protocol (also known as the Protocol on Environmental Protection to the Antarctic Treaty), which placed a moratorium on mining and drilling for oil for a minimum of 50 years. Unfortunately, despite the 50-year moratorium, future economic pressures will likely pose significant environmental threats (Panel 7.2) to the continent. Given this impending scenario, Antarctica's serenely primitive wilderness faces an uncertain future as debate continues over the question of tapping into the continent's wealth of mineral resources.

When it comes to Arctic exploration, the situation is little better. Offshore drilling started in the 1970s, but due to technical difficulties, production has been limited.

Panel 7.2. *Bahia Paraiso*

On January 28th, 1989, the *Bahia Paraiso*, an Argentine transport ship hauling supplies and tourists, ran aground 3 km off the Antarctic coast. Although no one aboard was injured, the wreck proved to be a disaster for the nearby coastal ecosystem, as a 10-m gash in the ship's hull released 950,000 l of diesel fuel into the ocean. The effects from the fuel spill on the local flora and fauna were mostly limited to sea bird, krill, and moss, with few populations suffering mortality rates greater than 20%. Because the spill was the first reported incident of its kind in the Antarctic region, it alarmed environmental groups, which viewed the disaster as a foreshadowing of future accidents if trends in tourism and ship transport were to continue. Two months later, the *Exxon Valdez* oil spill in Alaska's Prince William Sound sent an even stronger message to environmental organizations for the need to protect Antarctica's unique environment from similar accidents. The Madrid Protocol, which is viewed as a model for future environmental treaties, was largely a result of the response to the two incidents.

But, with renewed interest in the Arctic, the oil and gas industry is designing and building new classes of modern drill-ships and tankers for the Arctic, able to penetrate deeper waters and reservoirs. Submersible drilling rigs can be used in the Arctic, but they work best in water depths of less than 30 m, which is not much help when most of the new prospects in the Arctic are at depths of up to 1,000 m! Despite the problems, the race is on to start offshore drilling in the Arctic and it is open to any and all who can get there and develop it.

With energy demand forecast to increase 58% over the next 25 years, Arctic and Antarctic reserves will be rapidly exhausted. Clearly, what is needed is a renewable energy source. One such source is geothermal energy, which is the heat within the Earth and can be "mined" by extracting hot water or steam, either to run a turbine for the generation of electricity or for direct use of the heat itself. It is an energy source that is rapidly gaining steam, with everyone from Google to Middle Eastern oil sheiks lining up to get a piece of the action. For example, Google has already invested millions in enhanced geothermal systems, while Masdar, a clean-tech city in Abu Dhabi, has awarded multi-million-dollar contracts to begin drilling geothermal wells.

GEOTHERMAL ENERGY

The attraction of renewable energy sources is that nature continuously replenishes them and they have a smaller carbon footprint than fossil fuels. One example of an untapped renewable energy source is the hydrothermal vent, the same underwater feature that produces SMS. In addition to producing SMS deposits, hydrothermal vents are natural geysers of magma-superheated water (Panel 7.3), which can be used for energy. Found atop mid-ocean ridges at an average depth of 2,300 m, hydrothermal and geothermal locations are found along tens of thousands of kilometers of mid-ocean ridges that have yet to be explored.

Panel 7.3. Superheated water

Tectonic plate movement opens fissures on the sea floor into which seawater is forced down by the high pressure of the water column above it. Magma superheats the seawater to a temperature as high as 400°C, creating a geyser with a velocity of 1–5 m per second (mps). Because the pressures at depth are up to 3,200 psi, the hot fluid stream does not flash into steam.

The sizes of the vents that have been surveyed are impressive. For example, the Juan de Fuca Ridge 320 km off the coast of Seattle has an active vent field over 150 m wide and 300 m long. Within that field are more than 15 vents up to 25 m in diameter that continuously eject water at 370°C at a rate of 3–5 mps. Each of the vents theoretically offers 40 billion watts' equivalent of thermal energy.

While 2,300 m is deep, it is not beyond human reach, so tapping these volcanic vents will not be any more technically challenging than drilling for oil, although it will probably be more difficult to capture hydrothermal energy and convert it economically to electrical energy. However, with the cost of all kinds of energy rising, the appeal of undersea geothermal energy will undoubtedly grow. To tap the energy generated by these hydrothermal vents, engineers will probably bring the high-temperature liquid to the surface via a network of insulated pipes to an offshore power plant on a platform similar to an oil rig. The liquid would rise by a combination of vent flow velocity, convection, conduction, and flash steam pressure generated as the superheated fluid rises and the ambient pressure diminishes. Since harnessing hydrothermal energy from vents on the ocean floor might represent the ultimate renewable source, the race will be on to find these reservoirs of energy. Since it is estimated that the thermal output of all the known vents in the world is about the same as current worldwide power-generating capacity, finding these vents will be a lucrative business. With tens of thousands of kilometers of ocean ridges yet to be explored, companies are already dispatching autonomous underwater vehicles (AUVs) to search (Figure 7.4).

AUTONOMOUS UNDERWATER VEHICLES

Much like the Martian rovers Spirit and Opportunity, AUVs are able to explore unaided. Almost entirely cut off from the world above, they can sniff out active

Figure 7.4. Woods Hole Oceanographic Institution's autonomous underwater vehicle *Sentry* is lowered into the North Atlantic for deep-ocean testing in April 2008. Photo by Chris German. Courtesy: Woods Hole Oceanographic Institution.

hydrothermal vents, sample what they find, and return to the surface, using cutting-edge sensing and navigational technologies. Despite their autonomous designation, AUV exploration is fraught with difficulty, not least because communicating with underwater vehicles is even harder than talking to robots on Mars. Since electromagnetic waves barely propagate through water, the best way to send data is via acoustic signals, but these weaken rapidly with distance. Up to a distance of 5 km, the operators have some idea where the AUV is but beyond that, they have to hope the vehicle's navigation system (Panel 7.4) will prevent the AUV from becoming lost.

An example of such an AUV is the Autosub6000, outfitted with forward-looking vertically scanning obstacle-detection sonar and improved terrain-following control software, giving the vehicle the ability to operate safely, close to the ocean floor. Equipped with the latest obstacle-avoidance systems, Autosub6000 can be deployed into hostile and rugged terrain without causing its operators too much concern. Another AUV example is the Autonomous Benthic Explorer (ABE), whose unusual shape renders it particularly suited to survey work on the ocean floor. The ABE (Figure 7.5) is already a veteran of hydrothermal vent expeditions, having been deployed on the South Atlantic Ridge in March 2005.

Panel 7.4. Underwater navigation methods

Since GPS does not work underwater, due to the attenuation of radio waves through the water, engineers must employ alternative methods for navigating AUVs. One method is to use acoustic long-baseline navigation (LBL), which requires positioning acoustic transponder beacons (ATBs) on the sea floor. This technique allows an AUV to measure the round-trip travel time of sound signals transmitted and received between it and the ATB. With two or more ATBs plus a measurement of AUV depth, the AUV's position can be calculated. A drawback of LBL is that it takes time to deploy and calibrate the ATB network and the ATBs have a finite working range within which the AUVs can acoustically hear.

Future AUV navigation will probably use "synchronous-clock one-way-travel-time navigation" – a new acoustic method for underwater navigation that aims to eliminate the need for deploying ATBs by turning the support ship into a moving beacon. By equipping each AUV with an acoustic modem (a device that allows the AUV to broadcast information) and by having very precise clocks on the AUV and support ship, it is possible to measure the one-way-travel-time (OWTT) from the surface ship to the AUV each time the surface ship broadcasts. Also, because it is a modem, the ship can also use it to send its GPS position along with the OWTT packet, meaning the AUV has a range to the ship and knows where the ship is according to GPS.

To help it detect hydrothermal vents, the ABE is outfitted with a myriad variety of sensors, including scanning and multibeam sonar, a magnetometer, a digital still camera, hydrographic sensors, and temperature probes. ABE operates autonomously from the vessel. It has no tether, is controlled in real time by onboard computers, and uses batteries for its power. After launch, the vehicle descends to the ocean floor using a descent weight and during its spiral trajectory descent, it uses acoustic long-baseline transponder navigation and acoustic Doppler measurements to determine its position and velocity. Once on the ocean floor, ABE performs a series of checks and releases its descent weight to become neutrally buoyant before beginning its survey. A typical dive usually consists of a hydrothermal plume survey at constant depth, following the sea floor at an altitude of 50–200 m, occasionally dropping to 5-m altitude to take digital photographs. One of the first signs ABE looks for when searching for a hydrothermal vent is turbulence in the seawater, generated by the water rising from the vent. From the perspective of the underwater robot, this sign is recognized as a patchy, irregular structure and background currents caused by the rising water. As the plume of water emitted from the vent rises, the AUV uses its optical backscatter (OBS) equipment to measure differences in temperature and salinity – both markers for hydrothermal vent activity. To check what it has discovered, the AUV may make two or three passes along the axis of the

Figure 7.5. The Autonomous Benthic Explorer (ABE). Although hydrothermal plumes can be detected up to a distance of several kilometers from the emitting vent field with standard in-situ sensors, turbulent and chemically altered seawater makes location vent sites challenging, which is why AUVs such as the ABE are deployed. Courtesy Woods Hole Oceanographic Institution.

ridge at a distance of about 1 km from the vent. Once the AUV thinks it has detected a vent, it will map the location at progressively lower altitudes, gradually building up a bathymetric map of the area (Figure 7.6) using its multibeam sonar to produce accurate bathymetric maps down to sub-meter scales. Thanks to this detail, the ABE has mapped tectonic features such as faults with great clarity, and resolved them into multiple components such as volcanic domes and hydrothermal vents. This bathymetry can then be used to reconstruct the tectonic history by providing sufficient detail and precision so that faults can be computationally removed to reveal the dome-like structure from which the rift evolved. ABE can also map and locate hydrothermal plumes by measuring heat fluxes from previously discovered hydrothermal vent fields. It does this by using instruments that measure temperature and salinity while following a tight grid pattern repeatedly above the vent field. The grid pattern (Figure 7.7) is also employed by the ABE to locate and characterize undiscovered hydrothermal vent sites. It starts with clues provided by towed systems that indicate a vent site and, by executing a sequence of grid patterns at increasingly finer scales and increasingly close to the sea floor, it generates a fine-scale bathymetric map of the vent and surrounding environment.

Figure 7.6. A bathymetric map generated by the ABE using multibeam sonar. Courtesy Woods Hole Oceanographic Institution.

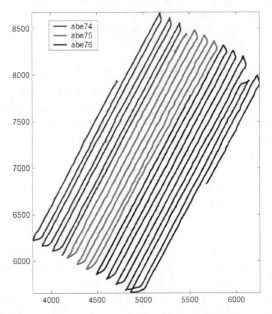

Figure 7.7. Grid pattern employed by the ABE to locate hydrothermal vents. Courtesy Woods Hole Oceanographic Institution.

Once it has positively identified a hydrothermally active region, it uses high-resolution digital photography to provide scientists with preliminary data for characterization of the site. Finally, after a 15–30-hr dive, the ABE simply releases the ascent weight to become positively buoyant and returns to the surface.

THE FUTURE

On August 2nd, 2007, Russia dropped a titanium capsule bearing its flag onto the Arctic floor, highlighting its bid for a price of underwater real estate that may contain billions of dollars in untapped energy. The brazen act prompted other nations, including the US, Canada, Denmark, and Norway, to make competing claims. Never before has the world's attention been so fixed on the deep ocean. Inflated oil, mineral, and gas prices are pushing industries into remote seas once too expensive to tap. Ironically, this extraordinary geopolitical and economic interest in the deep ocean is happening at a time when we still know so little about the nature of the ocean (maps of Mars are about 250 orders of magnitude better than maps of the ocean floor). What *is* certain is that deep-sea mining, while a new industry, has the potential to balloon as oceanographers and AUVs discover more and more mineral deposits on the ocean floor. Nautilus Minerals and Neptune Minerals, the world's first two deep-sea mining companies, have already launched operations, and India has announced a $100 million-per-year initiative to probe farther into its own cobalt and manganese-rich waters. While the mining of minerals will prove lucrative, the real hotspots will be hydrothermal vents.

Since research and development in the fields of deep-sea mining and energy exploitation have the potential to influence the industrial future and technological potential of industrialized countries, it is inevitable these endeavors will go ahead, regardless of any treaties. Those concerned about the environmental effects of exploiting the pristine wilderness of the Arctic Ocean and the Antarctic Sea can take comfort in the fact that the future of deep-ocean mining and energy exploitation will be inextricably tied to all deep-sea ventures, requiring the participation of oceanographers and highly skilled engineers. By the same token, advances in our understanding of the ocean are being generated by both those who seek to study the oceans and those who seek to exploit it. For example, geologists mapping the contours of the Arctic Ocean floor for territorial claims have provided some of the first images of this remote polar terrain.

However, energy exploitation is not the only component of ocean industry that is growing. The deep ocean environment is also the frontier for the search for new medicines, and it is this subject that is explored in the next chapter.

REFERENCES

[1] Aarli, J.A.; Vaernes, R.; Brubakk, A.O.; Nyland, H.; Skeidsvoll, H.; Tonjum, S. Central Nervous Dysfunction Associated with Deepsea Diving. *Acta Neurologica Scandinavica*, **71**, 2–10 (1985).

[2] Todnem, K.; Nyland, H.; Dick, A.P.K.; et al. Immediate Neurological Effects of Diving to a Depth of 360 Metres. *Acta Neurologica Scandinavica*, **80**, 333–340 (1989).

[3] Vernes, R.J.; Klove, H.; Ellertsen, B. Neuropsychologic Effects of Saturation Diving. *Undersea Biomedical Research*, **16**, 233–251 (1989).

8

Ocean Medicine

In the field of pharmaceutical development, there is a problem. Nearly half of all new drugs are derived from terrestrial microorganisms, but because the soil has been exploited to the limit, there are few drugs left to discover. This has forced pharmaceutical companies to abandon many of their soil-based research and development programs and search for new drugs in other locations, such as the ocean. The good news is that the prospect of finding a new drug in the ocean, especially among coral reef species, may be 300–400 times more likely than isolating one from a terrestrial ecosystem. Although terrestrial organisms exhibit great species diversity, marine organisms have greater *phylogenetic* diversity. In fact, thousands of marine species have never been studied and the potential of these marine organisms as sources of medicines has created a renaissance of interest in exploring the deep ocean. The work of searching for marine drugs is carried out by bioprospectors, who spend much of their time diving hundreds and sometimes thousands of meters underwater in search of what they hope will be the next miracle medicine. Just a few years ago, these scientists strapped on scuba gear and hand-scooped samples from the ocean floor before taking it back to the laboratory for analysis. Today, the pioneering researchers have been joined by other underwater bioprospectors who work in submersibles in deep-water sites from the South Pacific to Hawaii.

This new cadre of scientists work in the rapidly emerging field of marine biotechnology – a discipline encompassing marine biomedicine, materials technology, bioremediation, molecular genetics, genomics, *and* bioinformatics. The work is not easy because the identification and extraction of natural products require major search and collection efforts that sometimes require the collection of hundreds of kilograms of an individual species in the hopes of identifying a single useful compound. For example, one bioprospecting group collected 1,600 kg of a sea hare to isolate just 10 mg of a compound used to fight melanoma, while another group collected 2,400 kg of an Indo-Pacific sponge to produce only 1 mg of an anti-cancer compound. If that sounds like a lot, consider this: as much as 1 kg of a bioactive metabolite may be required for drug development! To some people, it might sound like a lot of work for little gain, but for prospecting scientists, there is always the possibility that the world's next miracle medicine may be found in the ocean, so they continue their search.

The driving force behind marine biotechnology is the enormous biodiversity and genetic uniqueness of life in the ocean. In 1998, in an effort to change how we look at this global biodiversity, scientists from the Scripps Institution of Oceanography (SIO) and departments at the University of California joined forces to establish the Center for Marine Biotechnology and Biomedicine (CMBB). CMBB's goal is to mobilize diverse scientific disciplines into a coordinated effort to explore new biotechnologies inspired by marine life. To date, CMBB and other agencies involved in marine biotechnology have identified more than 14,000 new chemical entities from marine sources [1], and more than 300 patents have been issued on marine natural products [2]. While marine biotechnology might be a new field, scientists have in fact been searching for ocean medicines for decades. The earliest known example of such exploration was the 1950s' discovery of spongothymidine and spongouridine from the sponge, *Tethya crypta*, which led to the development of antiviral, anti-cancer, and anti-HIV drugs [3]. Today, thanks to the help of CMBB and other agencies, a new generation of marine-derived pharmaceuticals is set to enter the market. Examples include a chronic pain treatment, Prialt (ziconitide), which has been isolated from the cone shell, *Conus magnus* [4], and the anti-cancer drug, Yondelis (Ecteinascidin 743), from the ascidian, *Ecteinascidia turbinata* [5].

DR. FENICAL

One of the scientists leading the way is CMBB's director, Dr. Bill Fenical. Dr. Fenical first became excited about the ocean at the age of 12 while on a trip to Florida. Not long after, he and his family moved to California, where Bill continued to enjoy the ocean by scuba-diving. Later, after gaining his PhD from the University of California, he sought to blend his love of the ocean with his career, and found a job as an Assistant Research Chemist at Scripps in 1973 – a position that focused on marine chemical ecology. By pursuing his research, Dr. Fenical became interested in the medical potential of the oceans. Today, his studies focus on the discovery of medicinally valuable compounds derived from marine microorganisms collected in tropical locations, as well as extreme environments such as arctic waters. His first success occurred in 1983, when Dr. Fenical discovered a sea fan off the coast of the Bahamas. He isolated a substance from the sea fan which acted as an anti-inflammatory in humans. Pseudopterosin, as the substance is known, was patented and the rights sold to Estee Lauder for use in skin-care products. Ten years later, pseudopterosin became one of the University of California's leading patent royalty earners and today the substance continues to be studied as a possible treatment for wounds:

> "Bill Fenical is a true pioneer of marine natural products chemistry, one of the first to see the great potential to benefit humanity from the study of marine organisms, particularly in the area of drug discovery. As the rate of discovery of new drugs from land organisms slows down, our marine resources will become ever more critical and valuable, something Fenical has understood for

over 30 years. Bill is also a wonderful communicator of his science, earning the respect of specialists and laypersons equally."

Charles Kennel, Director of Scripps

While he may never attain the status of Bill Gates, Steve Jobs, or the other technological innovators who routinely transform our daily life, Dr. Fenical is a pioneer in the same mold thanks to his pursuit of a discovery that has the potential to revolutionize medicine. Dr. Fenical is searching for a cure for cancer, and he is searching for it in the birthplace of life itself: the ocean. To date, Dr. Fenical and his team of researchers at Scripps have discovered and identified 15 genera of organisms, and they have barely begun. Already, two compounds derived from marine organisms that his team have discovered are undergoing human trials. The first, SAlA, is an anti-cancer drug being tested on volunteer cancer patients for its potential to combat malignancy of the bone, while the second, known as NPI-0058, may have the potential to reduce tumors in the lung. An author of more than 300 scientific papers, Dr. Fenical also helped found Nereus Pharmaceuticals, a private company licensed to bring medical discoveries to market. Using a 13-m trawler, *Osprey*, he takes his students on regular research trips in the Sea of Cortez in search of marine organisms he hopes will help him in his fight against cancer. During these trips, students don scuba gear twice a day and search the ocean floor for marine resources that eventually end up in their lab for analysis.

CMBB

In addition to their in-house research, CMBB researchers participate in cooperative programs with other universities, academic collaborators, and several pharmaceutical companies to develop new drugs from the ocean. Recently, Scripps scientists isolated a chemical from a rare species of coral that shows promise as a potential drug to fight ovarian cancer, while other Scripps scientists discovered a chemical from a marine sponge that may be used to treat inflammation without the problems associated with aspirin. The marine sponge chemical is currently being used by more than 20 companies as a tool for understanding the inflammation process.

The work of marine bioprospecting is not limited to drug discoveries, however. In the emerging field of marine viral genomics, Scripps scientists also hunt for viruses. It is estimated that for every liter of ocean water, there are about 1 billion bacteria and 10 billion viruses, a number that easily makes viruses the most common ocean "predator", but until Scripps scientists started their work, ocean viruses were a mystery. Through their research, Scripps scientists are slowly shedding light on this marine enigma and recently published the first scientific paper reporting the full DNA sequence of a marine virus.

Scripps is also making advances in the field of genetics by transferring genetic material from one organism to another in an attempt to develop new cutting-edge pharmaceuticals – work that was fuelled by observations that some marine pharmaceuticals are produced by symbiotic microorganisms. However, since many of these microorganisms cannot be cultured to produce sufficient quantities of the required pharmaceutical compounds, Scripps researchers are searching for genes that produce the desired compounds and to transfer those genes to a bacterium that is easy to culture. Currently, scientists are employing these techniques in the study of the anti-cancer agent bryostatin and the antifungal agent theopalauamide.

On another front, scientists are investigating bioluminescence, which has numerous applications in cell and molecular biology, biotechnology, and medical diagnosis. For example, luminescent bacteria can be used to detect contaminants in wastewater and their cloned genes provide a tool for monitoring gene expression. One of the most common sources of marine bioluminescence is the single-celled luminescent plankton known as dinoflagellates, which are very sensitive to water motion and can be used as a tool for visualizing complex flow patterns in water.

DRUGS TO MARKET

There is no doubt the ocean is an exceptional source of bioactive natural products, many of which exhibit features not found in terrestrial natural products, but why should this be? The simple answer is that marine organisms have not only evolved physiological and biochemical mechanisms that include the production of bioactive compounds for such purposes as communication but, more pointedly from the pharmaceutical industry's perspective, have also evolved protection mechanisms from infection. But it is not just the physiological and biochemical diversity that is impressive. It is the *biological* diversity that makes the ocean a veritable drugstore since, among the 34 phyla[1] of life, only 17 occur on land, whereas 32 occur in the ocean. Given this diversity, it is not surprising that a scientist such as Dr. Fenical and organizations such as the CMBB have isolated so many marine natural products[2] (Table 8.1). Given the considerable impact that marine natural products have had upon medicine to date, and the potential they are sure to have in the future, it is worthwhile examining some of these therapeutic treasures from the deep, but before we do, it is important to emphasize that word *potential*, since there is no guarantee that any of the compounds discussed here will see daylight. That is because, on average, it takes more than a decade and several hundred million dollars to move a potential drug along the pipeline (Panel 8.1). Also, bear in mind that for every new compound granted Federal Drug Administration (FDA) approval, thousands of others are abandoned because they were found to be either clinically ineffective or unsafe at therapeutically effective dosages.

---

[1]   Phyla is the plural of phylum, which is the equivalent to the botanical term "division". Some of these phyla overlap.

[2]   Of these, 25% are from algae, 33% from sponges, 18% from coelenterates (sea fans and soft corals), and 24% from invertebrate phyla such as ascidians, opisthobranch mollusks (nudibranchs), echinoderms (starfish, sea cucumbers, etc.), and bryozoans (moss animals).

Table 8.1. Chemicals and biological materials isolated from marine organisms [1].

Application	Original source	Status
Pharmaceuticals		
Anti-cancer drug (non-Hodgkin's lymphoma)	Sponge, *Cryptotethya crypta*	Commercially available
Anti-cancer drug	Bryozoan, *Bugula neritina*	Phase II clinical trials
Anti-cancer drug (tumor-cell DNA disruptor)	Ascidian, *Ecteinascidia turbinata*	Phase III clinical trials
Anti-cancer drug	Ascidian, *Aplidium albicans*	Advanced preclinical trials
Anti-cancer drug	Gastropod, *Elysia rubefescens*	Advanced preclinical trials
Anti-cancer drug	Sponge, *Discodermia dissoluta*	Phase I clinical trials
Anti-cancer drug	Sponge, *Lissodendoryx* sp.	Advanced preclinical trials
Anti-inflammatory agent	Marine fungus	In development
Anti-tuberculosis agent	Gorgonian, *Pseudopterogorgia*	In development
Anti-HIV agent	Ascidian	In development
Anti-malarial agent	Sponge, *Cymbastela*	In development
Medical devices		
Orthopedic and cosmetic surgical implants	Coral, mollusk, echinoderm skeletons	Commercially available
Diagnostics		
Detection of endotoxins	Horseshoe crab	Commercially available
Nutritional supplements		
Polyunsaturated fatty acids used in food additives	Microalgae	Commercially available

Panel 8.1. Bringing drugs to market

A new marine-derived drug candidate will first be subject to years of preclinical research. If the results are encouraging, the compound may be evaluated for efficacy and toxic side effects using animal models and/or computerized simulations. If the drug continues to show potential, the research lab will search for a partner in the pharmaceutical industry interested in licensing the drug to pursue clinical development. The partner will then inform the FDA of its intentions to develop the drug by submitting an Investigational New Drug application (IND). This is when the real work begins, because the drug has to perform well during several years of human clinical trials, meeting clinical endpoints for safety and exhibiting acceptable absorption, digestion, metabolism, and elimination (ADME) criteria. If, at the end of these phases, the drug still shows potential, the industry partner will submit a New Drug Application (NDA) to the FDA. Finally, once the FDA approves of the drug, it may be brought to market, concurrent with ongoing post-clinical studies to verify the absence of unanticipated risks or side effects.

One of the most promising candidates derived from marine natural products is Bryostatin 1, a compound that has been licensed to Bristol-Mayers Squibb, which has completed Phase I clinical trials in the US. Extracted from a species of bryozoan,[3] *Bugula neritina*, Bryostatin 1 is under investigation as an anti-cancer and memory enhancement agent. Laboratory trials have been promising, showing the compound to act synergistically with other anti-cancer drugs and to have a potent anti-leukemic effect against lung, prostate, and non-Hodgkin's lymphoma tumor cells.

Equally promising are dolastatins. Dolostatins are compounds that have been extracted from the marine shell-less mollusk, *Dolabela auricularia*. The compounds were first isolated in 1987 [6]. Within the family, dolastatin-10 is under investigation for use as an anti-tumor agent due to its promising anti-proliferative actions. Laboratory studies have demonstrated the compound's ability to inhibit growth against small lung cell cancer cell lines [7] and it has been used with mixed success in the treatment of patients with sarcomas.

A drug similar to Dolostatin 10 is Discodermolide, discovered in 1987 by scientists with the Harbor Branch Division of Biomedical Marine Research. Isolated from the Bahamian deep-sea sponge, *Discodermia dissoluta*, Discodermolide is a promising candidate for treating certain cancers. It works by chemically stabilizing target cells and arresting them at a specific stage in the cell cycle, thereby halting cell division. In addition to anti-cancer properties, the drug possesses immunosuppressive properties. The pharmaceutical company Novartis Pharma AG recognized the potential of the drug and licensed it for commercial development in 1998. After preliminary laboratory tests, the drug continued to show promise in combating pancreatic cancer and other drug-resistant cancers.

Another potential cancer-killer is Ecteinascidin-743 (ET-743), currently being developed at 13 centers across the European Union (EU) and the US to treat people with soft-tissue sarcomas. Soft-tissue sarcomas are cancers of the supporting tissues such as muscle or other tissues that support, surround, and protect the organs of the body. While the cause of soft-tissue sarcoma is largely unknown, what is known is that soft-tissue sarcoma is a serious and often life-threatening condition. For those diagnosed with the early stages of the condition, surgery is the main choice of therapy. For larger sarcomas, radiotherapy is usually used as well as surgery. Fortunately, ET-743 may represent new hope for those suffering from this condition. Isolated from a small marine animal called a sea squirt (a problem is, it takes 1 tonne of the animals to isolate just 1 g of ET-743 and up to 5 g are needed for clinical trials) living on mangrove roots in the Caribbean Sea, ET-743 exerts anti-tumor activity via several mechanisms, making it difficult for the cells to multiply. Research indicates the drug may also keep tumor cells vulnerable to chemotherapy and may induce a decrease in the speed of cell cycle multiplication, which would eventually lead to cell cycle arrest, thereby preventing cancer cells growing. Unlike many new

---

[3]   Bryozoans are a phylum of aquatic invertebrate animals about 0.5 mm long, which sieve food particles out of the water using tentacles lined with cilia.

drugs that are still in the development phase, ET-743 is already being sold by Zeltia and Johnson and Johnson under the brand name Yondelis. Approved for use in Europe, Russia, and South Korea for the treatment of advanced soft-tissue sarcoma, the drug is also undergoing clinical trials for the treatment of breast, prostate, and pediatric sarcomas.

CORAL AND BONE GRAFTS

In addition to sea squirts and mollusks, scientists have recently focused their attention on marine coral, which has been heralded as a miracle application for bone grafts [8]. Bone grafts are implants that provide a framework into which the host bone can regenerate and heal [9]. The implant also provides a framework to support the new tissue, blood cells, and soft tissue as they grow to connect fractured bone segments. Among the myriad situations requiring a bone graft are fusions of the spine, fractures, gaps in bones caused by trauma, and revision joint surgery. In fact, there are so many situations in which bone grafts are required that they are second only to blood transfusions on the list of transplanted materials. For example, every year in the US, more than 500,000 bone graft procedures are performed at a cost that exceeds $2.5 billion per annum. In 90% of these procedures, either allograft or autograft tissue is used, the latter option being the current standard, in which tissue is harvested from the patient. Once harvested, the graft is placed at the injury site. Of course, harvesting the autograft requires additional surgery at the donor site – a procedure that can cause complications, such as inflammation and chronic pain that may outlast the pain of the original surgical procedure! Another problem is one of supply, because quantities of bone tissue that can be harvested are obviously limited. While allografts are used as an alternative to autografts (these are taken from donors or cadavers), their use presents risks, too, because although allograft tissue is treated by tissue freezing, freeze-drying, gamma irradiation, and a myriad other procedures, the risk of disease transmission (such as hepatitis B, for example) from donor to recipient is always present. Fortunately, a more effective and less risky alternative is available thanks to a product called hydroxyapatite (HA). This product can be produced by the exoskeletons of marine coral (Figure 8.1), and can act as a bone graft to facilitate bone re-growth. Because HA is similar in structure to human bone, the use of the product can fill voids caused by fractures.

Given the potential medical applications of marine coral, it was inevitable that sooner or later, the biotech industry would exploit its potential, which is why, today, many bone grafts are performed using Pro Osteon, a biotech synthetic material derived from marine coral. The material, sold by Interpore Cross International, facilitates the natural healing process without risking the major drawbacks of previous grafting methods (in 20,000 procedures using Pro Osteon, there has not been a single case of rejection). The material is made by subjecting the coral to a chemical process that converts it to HA, which has the same mineral content as human bone. After the conversion process, the porous, interconnected structure of the coral remains intact, providing an ideal matrix through which new bone tissue can grow.

Figure 8.1. Hydroxyapatite can be produced from the exoskeletons of marine coral and is used as a bone graft to facilitate bone growth. Courtesy Wikimedia.

At present, the tropical coral genera *Porites*, *Alveopora*, *Acropora*, and *Goniopora* are the most common ones used as bone substitutes, but scientists at the Max-Bergmann Center of Biomaterials at Dresden University of Technology in Germany have had success using bamboo corals (Figure 8.2), which are often found at depths of more than 1,000 m. These corals have jointed axes made of bony structures alternating with nodes made of a protein-based material called gorgonin, giving the coral an appearance that resembles fingers. Since the skeletal structure of bamboo coral is almost identical to that of bone, it will not be surprising if cultivation of bamboo corals opens a new path for the development of natural bone implants.

FUTURE BIOPROSPECTING

For centuries, people believed that life in the ocean ended about 100 m below the surface, which is the maximum depth sunlight penetrates. In recent years, however,

Figure 8.2. Bamboo coral possesses a skeletal structure almost identical to bone. These corals, found at depths of up to 1,000 m, are being studied as a substitute for bone. Courtesy NOAA.

advances in submersible technology have enabled scientists to explore the deep ocean, where they have discovered a veritable cornucopia of life that holds tremendous potential for human benefit. Already, more than 15,000 natural products have been discovered, and this number continues to grow. While bioprospecting and deep-ocean exploration are in their infancy, the novel biology of the organisms discovered to date and their potential for revolutionizing the medical realm mean scientific interest will be increasingly focused on realizing the potential that exists in the deep ocean. And, inevitably, as a growing body of scientific reaffirms that deep-sea biodiversity holds major promise for the treatment of human diseases, exploration will surely venture ever deeper in search of untapped resources.

REFERENCES

[1] Proksch, P.; Edrada-Ebel, R.A.; Ebel, R. Drugs from the Sea—Opportunities and Obstacles. *Marine Drugs*, **1**, 5–17 (2003).
[2] Kerr, R.G.; Kerr, S.S. Marine Natural Products as Therapeutic Agents. *Expert Opinion on Therapeutic Patents*, **9**, 1207–1222 (1999).
[3] Newman, D.J.; Cragg, G.M. Marine Natural Products and Related Compounds in Clinical and Advanced Preclinical Trials. *Journal of Natural Products*, **67**, 1216–1238 (2004).

[4] Olivera, B.M. ω-Conotoxin MVIIΛ: From Marine Snail Venom to Analgesic Drug. In: N. Fusetani (ed.), *Drugs from the Sea*, p. 74. Karger, Basel (2000).

[5] Rinehart, K.; Holt, T.G.; Fregeau, N.L.; Stroh, J.G.; Kiefer, P.A.; Sun, F.; Li, L.; Martin, D.G. Ecteinascidins 729, 743, 745, 759A, 759B, and 770: Potent Antitumor Agents from the Caribbean Tunicate *Ecteinascidia turbinata. Journal of Organic Chemistry*, **55**, 4512–4515 (1990).

[6] Pettit, G.R.; Kamano, Y.; Herald, C.L.; Tuinman, A.A.; Boettner, F.E.; Kizu, H.; Schmidt, J.M.; Baczynskyj, L.; Tomer, K.B.; Bontems, R.J. The Isolation and Structure of a Remarkable Marine Animal Antineoplastic Constituent: Dolastatin 10. *Journal of the American Chemical Society*, **109**, 6883–6885 (1987).

[7] Turner, T.; Jackson, W.H.; Pettit, G.R.; Wells, A.; Kraft, A.S. Treatment of Human Prostate Cancer Cells with Dolastatin-10, a Peptide Isolated from a Marine Shell-Less Mollusc. *Prostate*, **34**, 175–181 (1998).

[8] Vuola, J.T.; Böhling, T.; Kinnunen, J.; Hirvensalo, E.; Asko-Seljavaara, S. Natural Coral as Bone-Defect-Filling Material. *Journal of Biomedical Materials Research*, **51**(1), 117–122 (2000).

[9] Parikh, S.N. Bone Graft Substitutes: Past, Present, Future. *Journal of Postgraduate Medicine*, **48**(2), 142–148 (2002).

Section IV

Revolutionary Undersea Medicine

Can a man breathe like a fish? It may sound like a question asked by science fiction writers, but scientists are close to resolving the problem and have already filed patents for diving equipment that may soon allow divers to ascend and descend without the penalty of decompression stops. One way of achieving unencumbered underwater freedom is to breathe liquid – a technology dramatized in the science fiction movie, *The Abyss*. Another option is to utilize artificial gills, and a third may be to apply the technology of an advanced nanotechnological device known as the vasculoid. Here, in Section IV, the potential of liquid breathing is discussed and we meet an Israeli inventor who has developed a breathing apparatus that will allow breathing underwater without the assistance of compressed-air cylinders. Finally, in Chapter 10, we explore over-the-horizon technology that may have the potential to ultimately transform humans into an aquatic species.

9

Liquid Breathing and Artificial Gills

"It will happen. Surgery will affix a set of artificial gills to man's circulatory system – right here at the neck – which will permit him to breathe oxygen from the water like a fish. Then the lungs will be by-passed and he will be able to live and breathe in any depth for any amount of time without harm."

Jacques-Yves Cousteau

Liquid breathing is a type of respiration in which a human breathes an oxygen-rich liquid such as a *perfluorocarbon* (PFC).[1] PFCs were synthesized during the development of the atomic bomb (the Manhattan Project) when they were given the codename "Joe's stuff". PFCs are organic compounds in which all hydrogen atoms have been replaced by halogens, usually fluoride. In medical applications, the compounds are being evaluated as contrast agents for computerized tomography (CT) and magnetic resonance imaging (MRI), as sensitizing agents during radiotherapy, and as possible oxygen-carrying agents. Because PFCs are stable, inert compounds, they do not react with living tissues, which makes them an ideal candidate for all sorts of medical applications. Currently, the primary application of liquid breathing is the medical treatment of lung problems in babies born prematurely [1, 2]. These babies usually have underdeveloped lungs and because PFCs can carry more oxygen than air, liquid breathing may help relieve respiratory distress until the lungs are able to function with regular air. More recently, the procedure of filling lungs (in clinical use, the lungs are usually not filled and liquid ventilation is used in conjunction with conventional gas ventilation) with a PFC called a perflubron (Panel 9.1) has been applied to adults with acute respiratory failure and infants with acute respiratory distress syndrome (ARDS). The practice helps heal the lungs because the liquid encourages collapsed alveoli to open and provides a better exchange of oxygen and carbon dioxide (CO_2) for lungs that are

---

[1] PFC liquids have a quarter of the surface tension, 16 times the oxygen solubility, and three times the CO_2 solubility of water. Since oxygen and CO_2 dissolve so easily in this liquid, it is excellent for carrying oxygen.

not fully functional. While the procedure has been proven in the clinical setting, liquid breathing may also have potential applications for diving [3, 4].

Panel 9.1. Perflubrons

Alliance Pharmaceutical Corporation makes a perfluorooctyl bromide (perflubron), marketed under the name LiquiVent^{TM}. The product is used for partial liquid ventilation. The fluid is administered to the lungs in conjunction with mechanical ventilation in adults with acute respiratory failure [5–9] and babies [1, 10] born prematurely. The perflubron helps open the alveoli (air sacs) of the patients and facilitates gas exchange in the lungs.

WHY BREATHE LIQUID?

When diving using air or mixed gas, the pressure inside the lungs must equal the pressure outside the body, otherwise the lungs will collapse. Since the external and internal pressures must be equal, the required gas pressure increases with depth to match the increased external water pressure. Unfortunately, these high pressures, especially when released quickly, may cause significant physiological problems such as air emboli and decompression sickness (DCS). Diving mammals and freedivers who dive to great depths on a single breath have few problems with DCS despite a rapid ascent to the surface. The reason freedivers and freediving mammals are not troubled by DCS has nothing to do with the fact that a single breath of gas does not contain enough total nitrogen to cause tissue bubbles on decompression, but because those who freedive spend such a short period of time on the bottom. But the dangers of diving breathing air or mixed gas are not restricted solely to DCS. Pain due to expanding or contracting trapped gases can potentially lead to *barotrauma*, a potentially damaging condition that may occur during either ascent or descent but is most severe when gases are expanding. Then there is *dysbaric osteonecrosis*, which is manifested as bone lesions most commonly on the body's long bones. This chronic disease, often observed in commercial divers, is thought to be related to the evolution of gas bubbles that may or may not be diagnosed as DCS. As you can see, there are several hazards associated with breathing gas, so it makes sense that divers would search for an alternative, safer means that would permit them to stay down longer and deeper without suffering all these problems.

As we have seen, one solution is to use an atmospheric diving suit (ADS), but these are bulky, expensive, and require a surface support team. Another more moderate option to deal with narcosis is to breathe mixed gas such as heliox or trimix, or to use closed-circuit rebreathers (CCRs), as discussed in Chapter 2. However, these options still don't avoid the problem of bubbles and DCS, because helium still dissolves in tissues and causes bubbles when pressures are released, just like nitrogen. A third option may be liquid breathing.

Imagine a diver's lungs ventilated by liquid. The pressure inside the diver's lungs could accommodate changes in the pressure of the surrounding water without the huge gas partial pressure exposures required when the lungs are filled with gas. Furthermore, liquid breathing would not result in the saturation of body tissues with high-pressure helium, thereby removing the requirement for decompression. For those of you who may be scratching their heads, thinking liquid breathing sounds familiar, the technology was dramatized in *The Abyss*, a 1989 science fiction film in which Ed Harris's character, Bud, dives to crazy depths breathing a pink liquid. However, the history of liquid breathing goes back further than James Cameron's epic underwater thriller. In 1920, two researchers named Winternitz and Smith demonstrated that human lungs can tolerate large amounts of a saline solution without damaging them. Winternitz and Smith's work was tested again by Johannes Kylstra, who, in the 1960s, submerged mice in saline solution and demonstrated short-term survival [11]. In fact, it was Kylstra (Panel 9.2) and his colleagues who, by demonstrating that mammals could breathe a liquid medium, started a resurgence of research in the field of liquid breathing. In the 1960s, the concept of breathing liquid was studied to increase the escape depth from a submerged submarine. To test their theories, the investigators immersed mice in salt solutions. To allow sufficient oxygen to be dissolved in solution, the animals were subjected to increased pressures, in some cases up to 160 atmospheres, which is the pressure 1.6 km below the surface of the sea. However, the work of breathing was very taxing and most animals died within minutes of respiratory acidosis. Kylstra's work was followed by a breakthrough in 1966, when Clark and Gollan investigated the use of PFCs. To test their theory, Clark and Gollan [12] submerged mice and cats (although not together!) and watched as the animals breathed the liquid. After keeping the test subjects in the liquid for some time, the researchers returned them to normal breathing and the mice and cats appeared to suffer no ill effects.

For liquid breathing to work in humans, the fluid has to perform two functions extremely well: deliver oxygen to the lungs and remove CO_2. Obviously, air performs both functions very well, as do various combinations of diving gases such as helium and oxygen and hydrogen. The PFCs developed in recent years have been shown to dissolve oxygen and CO_2 very well. In fact, LiquiVent^{TM} has the ability to carry more than twice as much oxygen per unit volume than air. It is also inert, which means it is unlikely to damage sensitive lung tissue and it can be cleared from the lungs very quickly. So, despite all the promising research and despite what we see in the movies, why is this technique not ready for use in the diving arena? Unfortunately, despite all the wonderful benefits to divers, there is still one hurdle that liquid breathing must overcome and that is how to move the liquid in and out of the lungs. You see, it is much harder for human lungs to move liquid in and out than it is to breathe air, so even though PFCs are more effective carrying oxygen and CO_2, the advantage is lost if the liquid is not circulated rapidly. The problem in achieving this transport lies in the PFC's high viscosity and the corresponding reduction in its ability to remove CO_2. For liquid breathing to be used by divers, *total* liquid ventilation must be achieved. This means the liquid must carry away sufficient CO_2 *regardless* of depth, and this is where the problem lies because no matter how great

Panel 9.2. Snibby the dog

Experiments investigating gas exchange in saline-filled human lungs were first conducted by Johannes Kylstra in the 1970s, but before the procedure could be tested on humans, several tests were performed on animals (dogs were a favorite). The test dogs were usually studied in pressure chambers large enough to accommodate air-breathing investigators and their liquid-breathing experimental test subject. First, a tub was placed in the chamber and filled with saline solution, after which the dog was anesthetized, shaved, washed, tracheotomized, and given antibiotics before being suspended above the tub. An airtight cover was placed over the dog and was attached securely around the tub, after which the chamber was pressurized with air. Oxygen was then bubbled through the saline solution. Investigators were protected from oxygen poisoning by venting the space under the cover into the chamber exhaust. After bubbling of oxygen through the salt solution had stopped, the dog was lowered into the tub until completely submerged, after which the dog started breathing liquid (he didn't have much choice!). Liquid breathing was stopped by lifting the dog out of the tub and draining the liquid from the animal's airspaces through a hose, after which the dog's lungs were inflated forcefully with air.

The first mammal to survive liquid breathing was a friendly mongrel dog called Snibby. During his experience breathing liquid, Snibby's systolic and diastolic arterial blood pressures were a little lower than normal and his heart rate was also lower. His breathing rate was slower than normal, but this was to be expected because breathing liquid is hard work. Although his blood was saturated with oxygen, the carbon dioxide level increased markedly. After spending 24 min underwater, Snibby was resuscitated and went on to make a full recovery. He was later adopted by the crew of the submarine rescue vessel, *H.M. Cerberus*, and served as a mascot.

As for Kylstra, he continued to conduct groundbreaking liquid-breathing research – work that earned him US Department of Defense funding, a photo essay in *Life* magazine in 1967, and a Lockheed Martin Award for Science and Engineering from the Marine Technology Society in 1970. In 1989, he worked as a consultant for *The Abyss* director, James Cameron.

the pressure is, the partial CO_2 pressure available to dissolve CO_2 into the liquid can never exceed the pressure at which CO_2 exists in blood, which is about 40 mm of mercury [13]. At this pressure, most PFCs require a volume equivalent to 70 milliliters per kilogram of bodyweight (about 5 l for an adult male) to remove enough CO_2 for normal resting metabolism. The situation is compounded when the diver's metabolic activity is considered, since the more active the diver is, the higher the breathing rate and the greater the production of CO_2. In fact, the

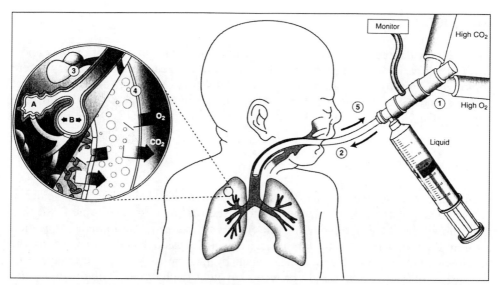

Figure 9.1. Illustration of perfluorocarbon liquid ventilation in a preterm infant. (1) The ventilator warms and oxygenates PFC liquid during slow instillation. (2) As liquid enters the side port of the endotracheal tube, the ventilator carries PFC to the distal areas of the lung. (3) As PFC liquid accumulates in the lungs, atelectatic regions of the lungs are expanded from A to B. (4) Oxygen and carbon dioxide are exchanged between alveolar PFC liquid and blood passing through the pulmonary capillaries. (5) Carbon dioxide is removed in expired gases by the ventilator. Courtesy American Association of Pediatrics.

lungs of a diver performing even light activities would be required to move 10 l or more of PFC every minute. This is a huge volume of fluid to move, especially when one considers PFCs are by several orders of magnitude more viscous than air and twice as viscous as water (in contrast, water is about 850 times the viscosity of air!).

So, how are PFCs used in hospitals and in research animals, you may ask? Well, in the research described earlier, a technique was employed that allowed the experimental animals to control the cycling of the respirator that circulates oxygenated liquid to and from the lungs – a method that reduced breathing effort by providing mechanical assistance (Figure 9.1).

Presently, there is no viable liquid breathing scuba system on the horizon. It is possible that future availability of enhanced biomedical-grade PFCs with diver-friendly physicochemical characteristics, combined with a system that reduces the work of breathing and removes CO_2, will be developed, but don't expect this equipment to be available in your local diving store any time soon. Nevertheless, it certainly is intriguing to think that a lung full of liquid could *prevent* a diver from drowning!

ARTIFICIAL GILLS

While it may be some time before divers breathe liquid, another invention may allow divers to breathe like a fish by simply extracting oxygen from the water. Before looking at how this invention might work, let's first ask why you and I cannot breathe underwater the way fish do. After all, fish need oxygen, just like us, and there is plenty of oxygen in the lakes and oceans, so there must be some difference.

Human ventilation system

In humans, the gas exchange surfaces are the lungs, which develop in the embryo from the gut wall – a development characteristic that relates us to some fossil fish! For those who studied biology at school, the structure of the human respiration system will be familiar. The larynx is a cartilage passageway connected to the trachea, which is a flexible tube held open by incomplete rings of cartilage. The trachea divides into left and right bronchi that enter the lungs and subdivide to form bronchioles, which are surrounded by circular smooth muscle fibers. At the ends of the bronchioles are groups of alveoli, which is where gas exchange occurs. The lungs possess typical features required by an efficient gas exchange system. First, they have a large surface area, thanks to about 600 million alveoli, which provide a surface area about the size of a doubles tennis court! Second, the single layer of flattened epithelial cells that comprise the alveoli ensure a short diffusion pathway. Third, a steep concentration gradient across the alveoli wall is maintained by blood flow on one side and air flow on the other side, which means oxygen can diffuse down its concentration gradient from the air to the blood, while, simultaneously, CO_2 can diffuse down its concentration gradient from the blood to the air. Finally, the moist surface of the alveoli provides a gas-permeable surface allowing gases to dissolve and then diffuse through the cells. Mechanically, the flow of air in and out of the alveoli comprises two stages: inspiration (inhalation) and expiration (exhalation). Since the lungs are not muscular, the thorax moves to facilitate ventilation thanks to the action of the intercostal muscles and the diaphragm.

Fish gills

In simple terms, fish exchange gases by indirect contact of blood with water in the gills. The mechanism by which the fish gill achieves this exchange has been studied in various scientific fields ranging from marine zoology to chemical engineering and it is this mechanism that holds the key to the development of an artificial gill. Research [14, 15] that has investigated the change in oxygen consumption with varying activity levels in fish has demonstrated there is a biological membrane that determines the rate at which oxygen is transferred from the water to the blood. However, before the oxygen is actually taken up by the blood, it must first pass through a structure known as the secondary lamellae. The secondary lamellae are very narrow channels that reduce gas transfer resistances in blood and water [16]. One way of thinking about this structure is as a gas exchange module for fish. By examining the structure of fish gills and the oxygen uptake mechanisms, scientists hope to one day create a gas-permeable membrane in an artificial gill in the same way as oxygen is taken up from water through a biological membrane in a biological gill. Biological gills are composed of thousands of filaments, which are, in turn, covered in feather-like lamellae [16]. The lamellae are only a few cells thick and contain blood capillaries. The structure provides a large surface area and a short distance for gas exchange. As the fish swims, inspired water from its mouth is routed to flow over the filaments and lamellae, and oxygen diffuses down a concentration gradient the short distance between water and blood, whilst CO_2 diffuses in the opposite direction, also down its concentration gradient. To maintain the concentration gradient, fish must ventilate their gills, which they do by continuously pumping water over them, expelling stale water behind [17]. If you were to look at the gill lamellae very closely, you would see they are arranged in a series of flat plates originating from the gill arch. On the upper and lower surfaces, there are several very thin vertical flaps that contain blood capillaries through which blood flows in the opposite direction to the flow of water over the gills. This mode of operation is called a *counter-current flow system* and it is a very effective diffusion pathway. This is because as the blood flows along and collects oxygen, it encounters water, which always has a greater oxygen content than itself, thereby ensuring the diffusion of oxygen into the blood will be maintained [18]. As the blood flows in the opposite direction to the water, it always flows next to water that has given up less of its oxygen, which means the blood is absorbing more and more oxygen as it moves along. Even when the blood reaches the end of the lamella, at which point it is 80% saturated, it is flowing past water that is at the beginning of the lamella and is more than 90% saturated. It is, quite simply, an extraordinarily efficient system that ensures that maximum possible gas exchange occurs.

Fish blood

In common with other vertebrate animals, fish blood consists of RBCs and plasma. The RBCs comprise leukocytes, thrombocytes, and erythrocytes, the latter being a round ellipse-shaped cell that contains dense hemoglobin. The molecular weight of fish hemoglobin is similar to mammalian hemoglobin, but the concentration of hemoglobin differs depending on the activity level of the fish. For example, active fish may have between 3 and 3.9 million erythrocytes per cubic millimeter compared with just 1.4–3 million in inactive fish. A comparison of hemoglobin saturations is shown in Table 9.1.

Table 9.1. Half saturation of hemoglobin for fish blood and human blood.

Species	Oxygen capacity (vol. %)	Temperature (K)
Cyprinus carpio (carp)	12.5	288
Scyliorhinus stellais (dogfish)	5.3	290
Human (male)	19.8	310

Oxygen consumption

Another function of the artificial gill is oxygen uptake. Not surprisingly, research has revealed that the oxygen consumption of fish varies with the activity level. It may sound obvious, but researchers needed to measure consumption rates in a biological gill before attempting to replicate the data in an artificial one. After much research, scientists created an artificial gill comprising two devices. One device was an oxygen uptake device that collected oxygen from the water to an oxygen carrier solution and the second was an oxygen release device that carried oxygen from the carrier solution to the air. In an attempt to replicate the conditions that exist in a biological gill, the oxygen carrier solution was cooled to 293 K – a temperature approximately the same as seawater and which also increases the oxygen affinity of the oxygen carrier solution, thereby enhancing oxygen uptake from the water to the oxygen carrier solution. In contrast, the oxygen release device was heated to 310 K to *decrease* the oxygen affinity of the oxygen carrier solution with the intent of enhancing the oxygen release from the oxygen carrier solution to the air. Much like a biological gill, the artificial gill extracts oxygen from the water to the oxygen carrier solution. Of course, while the biological gill achieves this by means of a biological membrane, the artificial gill uses a synthetic gas-permeable membrane, but the effect is similar [18, 19]. However, while the artificial gills function in a manner similar to a biological gill, scientists are unable to achieve the amount of oxygen required by a human.

You see, a diver requires much larger amounts of oxygen than fish because of their larger body volume. This causes problems for those designing artificial gills because a larger membrane surface is required to ensure a larger water flow rate (since it is the water that provides the oxygen). Despite the scientist's best attempts, the highest water flow rate achieved in an artificial gill is less than half of that in a biological one. The biological gill simply takes up oxygen much more effectively from water than an artificial one. One of the reasons for this performance difference is attributable to a large oxygen partial pressure difference between water and blood in the biological gill, which creates a greater driving force than can be achieved in the artificial gill [19]. Another reason is that the biological gill can take up oxygen more effectively at all water flow rates; nature is just more efficient. Scientists are working to improve the efficiency of the artificial gill, but they still have some work to do before they can match the performance of the biological equivalent. One modification the scientists are trying to implement into the artificial gill is to increase the oxygen partial pressure difference between the oxygen carrier solution

Figure 9.2. Alan Bodner, inventor of LikeAFish. Courtesy Alan Bodner.

and the air. If this can be achieved, oxygen release will be enhanced, but that is only part of the solution. To achieve a high oxygen partial pressure difference in the oxygen uptake, as is the case in the biological gill, a greater change in the oxygen affinity of the oxygen carrier solution is required. Once this is achieved, the artificial gill may begin to match the performance of its biological equivalent.

So, does this mean that divers are stuck with having to carry a bulky compressed-air cylinder on the back every time they want to dive? Not if Israeli inventor, Alan Bodner (Figure 9.2), has his way. Bodner has unveiled a novel approach to the artificial gill problem that will allow breathing underwater without the assistance of diving cylinders. Instead of utilizing a membrane gill, Bodner plans to use an industrial process for separating the gases from a liquid. His invention has already captured the interest of several major diving manufacturers as well as the Israeli Navy. If Bodner's invention is developed into a cylinder-free breathing system, it will revolutionize diving.

As we saw in Section I, there are several limitations to traditional scuba. First, the length of time a diver can remain underwater is related to the capacity of the compressed-air/gas cylinders. Second, divers are obviously dependent on compressed-air/gas facilities that are expensive to operate. Third, as any scuba-diver will tell you, the use of compressed-air/gas underwater causes all sorts of problems because as the cylinders are used up, they upset the balance of the diver. In other environments in which people have to rely on life support systems, engineers have attempted to overcome these limitations. For example, in nuclear submarines and onboard the International Space Station (ISS), life support systems generate oxygen from water by *electrolysis*, which is the chemical separation of oxygen from hydrogen. These systems require large amounts of energy to operate, which means they could not be adapted for scuba-diving. To overcome this limitation, Bodner decided to apply a mechanism used by fish. Instead of chemically separating oxygen from water, fish use the dissolved air that exists in the water to breathe. Now, you might be thinking there cannot be much air in water, but studies have shown that even at depths of 200 m, about 1.5% of the volume of water is air. It is not much, but it is enough to allow fish to breathe comfortably. Bodner's idea was simple: he would create an artificial system that would mimic the way fish use air in the water. Called "LikeAFish", Bodner's battery-powered artificial gill system was developed based on the principles of a well known physics law that describes the absorption of gas in liquids. Henry's Law states that the amount of gas that can be dissolved in a liquid is proportional to the pressure on that liquid. This means that as pressure is released, more gas will be forced out of the liquid and vice versa. In Bodner's prototype (Figure 9.3), pressure is lowered by means of a centrifuge that spins at a rapid rate, thereby reducing pressure inside a small sealed chamber that contains seawater. As the pressure is reduced, dissolved air escapes back into a gaseous state much like CO_2 is released from a can of Coke when it is opened. The air that is liberated as a result of this reduction in pressure is then transferred to an airbag for use by the diver. Because every liter of seawater contains 1.5% of dissolved air, the system must circulate about 200 l of water per minute to provide the breathing requirements of a diver. If realized (patents for the invention have already been granted in the US and Europe), divers would no longer be restricted by the amount of air/gas that can be carried in a cylinder, but simply by the amount of power available in a lithium battery.

Bodner expects to field an operational prototype within the next 2 years and although the popular press has voiced some concerns about the revolutionary technology, these issues are often based on a lack of understanding of what Bodner's device is capable of. For example, one problem cited was based on Bodner's assumption that a CCR diver will use only 1 liter of oxygen per minute. Such a rate

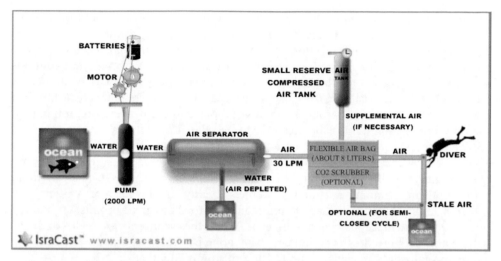

Figure 9.3. Alan Bodner's LikeAFish system. Convinced there was a better way to breathe underwater, Israeli inventor, Alan Bodner, decided to take advantage of a well known physical law called Henry's Law, which states that the amount of gas that can be dissolved in a liquid body is proportional to the pressure on the liquid body. The law works in both directions, so lowering the pressure releases more gas out of a liquid. In Bodner's invention, this is achieved by a very small centrifuge, which rotates rapidly thus creating under pressure inside a small sealed chamber containing sea water. To see a flow diagram of this image, go to *www.isracast.com/article.aspx?id=63* and to visit Bodner's site, go to *www.likeafish.biz*. Courtesy Alan Bodner.

of oxygen consumption (based on an average adult male engaged in light underwater activity) does not equate to consumption rates observed during heavy work such as swimming against a strong current, which can easily push consumption beyond 3 l per minute. However, as Bodner points out, he never made such an assumption and the figure of 1 l per minute is simply an industry average. To manage consumption, the device may be equipped with a heart monitor and a biosensor, which would increase the flow of water to keep pace with the demands of the diver's body.

Another perceived limitation expounded in the popular press was the assumption that the device was limited to only 200 l per minute – a rate that would be required to cope with higher oxygen consumption rates. The truth is that Bodner's invention has no such restriction. A third mistake made by the popular press science scribes was their concern on the issue of generating the high flow rates. To move 200 l per minute would require a greater centrifuge capacity and therefore require faster spin speeds. It was theorized that such a system might cause progressional instability problems that the diver would be unable to counter. Progression is the phenomenon of spinning objects to be tugged off center in the plane of the object's spin. The principle explains why bicycle wheels are prevented from wobbling while in motion. But on the back of a diver, the popular press writers argued, the spinning centrifuge would constantly be tugging gently at the diver and forcing him from where he would want to be. In reality, the problem of progressional instability is almost non-existent, since, even when operating at 200 l a minute, the device would only move the diver 1 m a minute. Hardly cause for concern. Finally, a question mark was raised regarding the concentration of oxygen in water. Although most water will contain 1.5% air, what happens if the diver encounters a "dead zone" that contains less dissolved oxygen, or if the water the diver is in is heavily polluted? Well, there is a reason the device is called LikeAFish. It is designed for waters that can sustain fish life, so hypoxic waters are probably not a good place to use it; therefore, ultimately, the only minor limitation of Bodner's device is related to location.

The potential of liquid breathing and artificial gills has intrigued generations of divers. While the very idea of breathing a liquid goes against everything we conceptualize about air-breathing humans, the successes in the science of liquid ventilation are stunning and unmistakable. Equally, in the field of artificial gill research, pioneers such as Alan Bodner continue to show great promise in making unconventional concepts seem mainstream. When these concepts are realized, divers will be able to go to depths no humans have gone before. They will also be one step closer to realizing the science fiction vision of homo aquaticus.

REFERENCES

[1] Davies, M.W.; Sargent, P.H. Partial Liquid Ventilation for the Prevention of Mortality and Morbidity in Paediatric Acute Lung Injury and Acute Respiratory Distress Syndrome. *Cochrane Database of Systematic Reviews*, CD003845 (2004).

[2] Leach, C.L.; Greenspan, J.S.; Rubenstein, D. et al. Partial Liquid Ventilation with Perflubron in Premature Infants with Severe Respiratory Distress Syndrome. *New England Journal of Medicine*, **335**, 761 (1996).

[3] Norris, M.K.; Fuhrman, B.P.; Leach, C.L. Liquid Ventilation: It's Not Science Fiction Anymore. *American Association of Critical-Care Nurses, Clinical Issues in Critical-Care Nursing*, **5**, 246 (1994).

[4] Shaffer, T.H. A Brief Review: Liquid Ventilation. *Undersea Biomedical Research*, **14**, 169 (1987).

[5] Hirschl, R.B. Advances in the Management of Respiratory Failure: Liquid Ventilation in the Setting of Respiratory Failure. *ASAIO Journal*, **42**, 209 (1996).

[6] Hirschl, R.B.; Pranikoff, T.; Wise, C. et al. Initial Experience with Partial Liquid Ventilation in Adult Patients with the Acute Respiratory Distress Syndrome. *Journal of the American Medical Association*, **275**, 383 (1996).

[7] Kacmarek, R.M.; Wiedemann, H.P.; Lavin, P.T. et al. Partial Liquid Ventilation in Adult Patients with Acute Respiratory Distress Syndrome. *American Journal of Respiratory and Critical Care Medicine*, **173**, 882 (2006).

[8] Kirmse, M.; Fujino, Y.; Hess, D.; Kacmarek, R.M. Positive End-Expiratory Pressure Improves Gas Exchange and Pulmonary Mechanics during Partial Liquid Ventilation. *American Journal of Respiratory and Critical Care Medicine*, **158**, 1550 (1998).

[9] Reickert, C.; Pranikoff, T.; Overbeck, M. et al. The Pulmonary and Systemic Distribution and Elimination of Perflubron from Adult Patients Treated with Partial Liquid Ventilation. *Chest*, **119**, 515 (2001).

[10] Shaffer, T.H.; Wolfson, M.R.; Clark, L.C., Jr. Liquid Ventilation. *Pediatric Pulmonology*, **14**, 102 (1992).

[11] Kylstra, J.A.; Tissing, M.O.; van der Maen, A. Of Mice as Fish. *Transactions, American Society for Artificial Internal Organs*, **8**, 378 (1962).

[12] Clark, L.C., Jr; Gollan, F. Survival of Mammals Breathing Organic Liquids Equilibrated with Oxygen at Atmospheric Pressure. *Science*, **152**, 1755 (1966).

[13] Cox, C.A.; Wolfson, M.R.; Shaffer, T.H. Liquid Ventilation: A Comprehensive Overview. *Neonatal Network*, **15**, 31 (1996).

[14] Hughes, G.M. How a Fish Extracts Oxygen from Water. *New Scientist*, **247**, 346–348 (1961).

[15] Hughes, G.M.; Morgan, M. The Structure of Fish Gills in Relation to their Respiratory Function. *Biological Review*, **48**, 419–475 (1973).

[16] Bijtel, J.H. The Structure and the Mechanism of Movements of the Gill Filaments in Teleostei. *Archives Néerlandaises de Zoologie*, **8**, 267–288 (1949).

[17] Eckert, R.; Randall, D.; Augustine, G. Exchange of Gas. In: R. Eckert, and D. Randall (eds), *Animal Physiology Mechanism and Adaptations*, 3rd edn, pp. 474–519. W.H. Freeman and Company, New York (1988).

[18] Matsuda, N.; Sakai, K. Technical Evaluation of Oxygen Transfer Rates of Fish Gills and Artificial Gills. *ASAIO Journal*, **45**, 293–298 (1999).

[19] Matsuda, N. Modeling of Gas Transfer in Biological and Artificial Membrane Modules and Acceleration of Gas Transfer Rate, PhD thesis of Waseda University (2001).

10

Becoming Homo Aquaticus

Beck:	They found this?
Doc:	No. I think they isolated him in genetic engineering. Homo aquaticus ....
Beck:	Was there such a thing?
Doc:	Danakil man? ... one of our ancestors. Named for the Danakil Alps in Ethiopia.
Beck:	Not exactly a big ocean country.
Doc:	It was when it counted a couple of million years ago when the seas came in and drove us into the water. Most homo sapiens didn't make it ... Danakil man ... adapted.

Excerpt from the script for the 1989 film, *Leviathan*

In *Leviathan*,[1] Tri-Oceanic Corp hires a crew of undersea miners for a 90-day mining operation. While exploring, one of the crew stumbles upon a wrecked ship, subsequently identified as the *Leviathan*. The crew opens a safe from the *Leviathan* and finds several records relating to deceased crew members and a bottle of vodka. Sixpack, one of the crew, hides the vodka for his own use but is persuaded by Bowman, another crewmember, to share it. The next day, Sixpack awakes feeling sick, with lesions all over his back. The crew is unable to offer any explanation, so they ask the computer, which suggests the lesions are the result of genetic alteration. Sixpack dies soon after, prompting the doctor to perform medical checks to make sure no one else is affected, but the doctor does not have the chance to examine Bowman. Sure enough, Bowman begins to feel the same effects as Sixpack did. Making matters worse, Bowman stumbles upon Sixpack's corpse and witnesses it mutating right in front of her. Unwilling to face the prospect of dying like Sixpack, Bowman commits suicide and her body is taken to sickbay, where Sixpack's body is

---

[1] Although the movie (which flopped at the box office) stole many ideas from John Carpenter's *The Thing*, it is recommended for those who enjoy "creature feature" movies and B-list films.

mutating. The reference to homo aquaticus is "Doc" suggesting the *Leviathan* was the location for experiments aimed at creating an aquatic human. The reference to Danakil man is a reference to the Aquatic Ape Theory (AAT), which suggests that many of the notable features of human physiology, though rare or even unique among land mammals, are common in aquatic ones.

While adaptation of the human body may be speculative and grounded more in science fiction than science, eventually, we will have the technology required to make the radical changes necessary to create a homo aquaticus. We may create this new breed of human using genetically engineered cells to produce the necessary tissue or organs. Alternatively, DNA could be cloned, the genes altered, and stem cells could be manipulated and implanted, possibly with the help of nanorobots. It may sound speculative, but such an idea is far from new. In 1957, noted science fiction author, James Blish, wrote a collection of stories that were published in *Seedling Stars*. The stories introduced a new word, "Pantropy", meaning "to change all", which described the idea of changing the human form to survive on other planet. It is a concept that may be realized by applying the relatively new technology of bioengineering.

BIOENGINEERING

There are a few problems to solve before humans can become homo aquaticus. First, we have to decide how the new human species will breathe. Once that problem is solved, it would be desirable to design a species that can exist both in and out of water. To do this, scientists could probably learn something from the various animals that have the ability to use both a gill and a lung system that allows them to breathe airborne and waterborne oxygen. Once those problems have been resolved, it will be necessary to find some way of protecting this new species from decompression sickness (DCS). And what about the skin? Everyone knows what happens to the skin after spending several hours in water, so there would have to be some way of increasing the skin's tolerance to saturation. Once these design issues have been addressed, scientists will probably want to offer minor options such as flipper feet and hands, eye protection, communication, and temperature tolerance. Obviously, a large proportion of the human organism will need to be redesigned. How will we do it?

First, we know we cannot breathe underwater because our lungs do not have the capacity to extract enough oxygen from water, but we can breathe whilst submerged in perfluorocarbons (PFCs). So we know the lungs can contain fluid and remain functioning and we know that by using PFCs, we can dive deeper for longer, but using PFCs will require extra equipment, so this really is not the answer to creating homo aquaticus.

But babies breathe liquid, don't they, so why can't we just use whatever technique they use? Well, babies do have a natural affinity with water, but contrary to popular belief, they do not breathe liquid in the womb. During gestation, the fetus survives in fluid, but oxygen/carbon dioxide (CO_2) exchange is provided by the mother via the umbilical cord. Babies know not to breathe underwater thanks to an instinct known as the mammalian diving reflex (MDR). What has the MDR got to do with becoming homo aquaticus, you may ask? Well, it is possible that the basic functions for breathing underwater already exist and by manipulating the system genetically, we could create a water-breathing human. Or perhaps it would be possible to employ bioengineering to modify our present breathing system by adding an external lung to extract the oxygen from the water and facilitate bimodal breathing. Alternatively, perhaps a human lung could be bioengineered to function like a fish gill? Many scientists have noticed the functional similarity between fish gills and mammalian lungs and have wondered whether a human version of such a lung could be developed to breathe water, assuming sufficient oxygen was present. Upon closer examination of this intriguing possibility, however, even the most optimistic scientist acknowledges that from a structural perspective, there are significant differences. In the case of the fish gill, the gas exchange is between water and blood, the structure of the gill consisting of a number of parallel planes in which gases are exchanged between water and blood by diffusion. From a bioengineering perspective, the system has the appearance of a radiator in which water flows directly past several capillaries that exist in each gill unit. It is this difference in geometry that presents the biggest problem to scientists hoping to engineer a homo aquaticus, since the lamellar structure of the fish gills act very differently from the spherical (alveoli) gas exchange units that humans must rely upon. The reason is simple physics. A hypothetical fish bioengineered with spherical gas units such as ours would not survive because gases diffuse much more slowly in a sphere than they do in a lamellar structure. If the opposite were to happen and water was introduced into a human lung, then the spherical shape of the alveoli would result in an increased diffusion time and would therefore adversely affect CO_2 elimination. In theory, the bioengineering solution to this problem is surprisingly easy, although a little more difficult in practice, since it would involve cutting a hole in the bottom of the lung! This may sound simple, but, in practice, the lung is not a single-space bag so this bioengineering feat would require that all several million alveoli be attached to a drainage plumbing system that could release water from the system. If this were achieved, it would enable exhalation to be performed through the base of the lung – a situation in which the lung could probably expel sufficient CO_2. By performing this rather elaborate surgery, the scientists would create a situation in which solution is pumped through the lungs instead of pumping into and out of the same section of lung. This would, in effect, create a similar situation to that which exists in a fish gill, since there would be a continuous fresh flow of water through the respiratory structure. It might be possible, but don't expect the surgery to be available for a while!

   Another adaptation that may be closer to the horizon is a bioengineered upgrade of the human body based on medical nanotechnology [1]. The "vasculoid", which is the name given to this particular upgrade, involves replacing a major part of the human body: the blood. It can be best thought of as an augmentation of the body and, if successful, would not only revolutionize the future of humans underwater, but also lay the foundation for some groundbreaking developments in the field of medicine.

Figure 10.1. Respirocyte in a blood vessel surrounded by red blood cells. The respirocyte is a nanobot capable of duplicating all thermal and biochemical transport functions of blood. © 2000 E-spaces and Robert A. Freitas Jr, 3danimation.e-spaces.com and www.rfreitas.com and Phillippe Van Nedervelde.

VASCULOID

One day, it may be possible for divers to breathe liquid or use artificial gills to extract oxygen from the seawater thanks to genetic engineering or by using applications such as Alan Bodner's invention. However, as revolutionary as these technologies may seem, a concept proposed by scientists, Robert Freitas and Christopher Phoenix, will make even these breakthroughs appear ordinary. Freitas and Phoenix's concept involves exchanging a person's blood with 500 trillion oxygen and nutrient-carrying nanobots. The system is called the vasculoid (a vascular-like machine) and it is designed to duplicate every function of blood [1, 2], albeit more efficiently (Figure 10.1).

The concept

One of the key elements of the vasculoid is the respirocyte, a type of nanobot that will be a key component of this artificial blood [3]. Freitas and Phoenix envisage each respirocyte will be constructed of 18 billion structural atoms precisely arranged to the last atom. Each respirocyte will have an onboard computer, powerplant, and molecular pumps capable of transporting oxygen and CO_2 molecules. The vasculoid would not only be capable of duplicating all the thermal and biochemical transport functions of blood, but it would also be able to perform these functions several hundreds of times more efficiently than biological blood. In essence, the vasculoid is nothing short of a mechanically engineered redesign of the human circulatory system. The complexity of the system is staggering, since it requires 500 trillion independently cooperating nanobots, yet the system weighs only 2 kg and is powered by nothing more than glucose and oxygen.

The key structural component of the vasculoid system is a two-dimensional vascular-surface-conforming array of 150 trillion square plates, called sapphiroids [3]. The sapphiroids are self-contained super-thin (only one micron thick) nanorobots that cover the entire surface of all blood vessels in the body, to one-plate thickness. Of the 150 trillion plates, 24 trillion are molecule-conveying docking bays where tankers containing molecules for distribution can dock and load or unload their cargo. Another feature of the array is the cellulock, of which there are 32.6 billion. At the cellulocks, boxcars carrying biological cells dock and load or unload their cargo. The remaining 125 trillion plates are reserved for special equipment and other applications. All the plates have watertight mechanical interfaces comprising metamorphic bumpers along the perimeter of each plate, which allow the bumper to expand and contract in area. It is a feature that permits the system to flex in response to body movements and to vascular changes.

At peak average male human metabolic rates, the body must transport more than 5 l of blood, which carries up to 125 cm^3 of essential molecules and other cellular elements required for metabolic function. In the vasculoid, these molecules and elements are containerized and distributed using subsystems such as the docking bays, cellulocks, and vasculocytes. One of the differences between the vasculoid and biological blood is the speed at which transport occurs. Biological blood moves at about 0.02–0.15 cm/sec, whereas the vasculoid moves at 1 cm/sec. The difference is thanks to the extraordinary efficiency of the vasculoid. You see, most of the molecules in regular blood are solvent in water, which is not necessary for human metabolism, whereas the vasculoid, thanks to sapphire-built tanker vessels, can store the necessary gases at a much higher concentration and are not diluted by water.

The efficiency of the vasculoid is perhaps best illustrated by the tanker fleet, which facilitate the transport of respiratory gases. Depending on activity level, the human body consumes between 1 and 20×10^{20} molecules of oxygen every second and generates a similar amount of CO_2 as waste product. The mass of all the red blood cells (RBCs) (about 2.4 l) can store 3.2×10^{22} oxygen molecules. But, because hemoglobin operates between 70 and 95% saturation, the active capacity of human blood is only 8.1×10^{21} molecules. To meet these targets, the vasculoid employs a respiratory fleet of between 1.68 (resting conditions) and 33.6 trillion tankers (high activity level). Thanks to their robust sapphire construction, each tanker is capable of holding gases at 1,000-atmospheres pressure, which means each unit can hold

9.48×10^9 molecules. This may seem an extraordinarily high pressure, but the tankers are operating way below their rupture pressure of 40,000 atmospheres [2].

Another function performed by the vasculoid is dealing with the excretion of waste water. Normally, this is achieved by the kidneys, which filter 18 gallons of blood every hour to cleanse the body of non-aqueous molecules. The kidneys accomplish this very efficiently, but with the vasculoid installed, only specific waste molecules are discharged, which means that water flow through the kidney and energy consumption in the organ are significantly reduced.

The vasculoid's function is not just limited to gas exchange and water excretion. Remember, this device replicates *all* the functions of the blood, one of which is glucose transport. During a typical day, a human metabolizes 3.08×10^{19} glucose molecules every second – a requirement that can rise dramatically, depending on exercise level. Once again, the vasculoid exceeds the design of the biological circulatory system; utilizing 17.6 trillion glucose tankers, the vasculoid is capable of storing 6.91×10^{22} glucose molecules, which is four times the number present in normal blood. Of those 17.6 trillion, only 1.1 trillion are required for resting metabolic requirements. The rest are held in reserve for periods of physical activity. The glucose tanker itself holds a cargo of 3.93×10^9 glucose molecules, but it does not just carry them; it also acts as a delivery system.

Another function of blood is the transport of proteins. The vasculoid does this too, transporting albumin, globulins (including antibodies), and other plasma proteins thanks to 0.12 trillion protein tankers. Similarly, lipids, including fatty acids and phospholipids, are transported in the vasculoid by 0.15 trillion lipid tankers, while just 0.03 trillion tankers take care of the transport of vitamins, trace elements, and specialty proteins. Regardless of what is being transported, there is sufficient tanker capacity to increase transport capacity to more than 100 times the recommended daily allowance if required. In fact, at maximum human metabolic rates, a total of 166.2 trillion (versus 11.78 trillion at rest) tankers would be engaged to transport all the necessary physiological molecules.

Of course, in addition to transporting molecules, the vasculoid must transport important cells such as white cells – a function that is achieved by utilizing cylindrical sapphire boxcars just 100 microns long and 6 microns in diameter. Because they are so small, the boxcars can easily pass through capillaries, meaning they have access to most human tissues. Embedded in the boxcar wall are computers and sensors that detect and interpret surface antigens; if the cells are hostile, they are simply refused transport and disposed of.

One physiological characteristic that has not been improved by the vasculoid is the role of the platelets (there are about 2 trillion of these in human blood). These have an important role in human physiology because they are storehouses for various molecules that affect vascular tone and they also play a vital role in the clotting process. In fact, in the event of a vessel breach, large amounts of extracellular fluid will leak if the breach is not promptly staunched.

But what happens to the regular blood, you may ask? Well, there are about 30 trillion RBCs present in human blood and these are superfluous as far as the vasculoid is concerned. That's right; regular blood will be removed and replaced with the more efficient vasculoid. In fact, because the vasculoid is so efficient, a fleet of only 34 million boxcars are required to meet the demands of regular blood.

Subsystems and concept of operations

The vasculoid's transportation function is achieved by a ciliary distribution subsystem (CDSS) comprising trillions of mechanical cilia that protrude from the vasculoid's sapphire-lined surface. In addition to facilitating transport speed, the CDSS assists the vasculocytes (remember, these are the mobile nanobots) in cleaning up after component malfunctions. However, its principal task is to transport the 166 trillion tankers and 32 billion boxcars through the physical circuit of the vasculoid. Just 110 nm long and 30 nm in diameter, each cilium is able to generate sufficient force to grapple and manipulate the various containers that are part of the cargo transport stream such as the tanker or boxcar [3].

Once the CDSS has done its job and transported the cargo to the appropriate sites, the containers must dock with the vasculoid surface and pass the delivered materials through the surface to the inside portion of the vasculoid. Once the cargo has made its way into the inside portion, the cargo diffuses into the tissues as required. The tankers, which carry molecules, offload their cargo in docking bays, whereas the boxcars, which carry cells, offload at cellulocks (Table 10.1).

Table 10.1. Tankers and boxcar cargo [3].

Tankers	Boxcars
Gas	Red cell
Water	White cell
Glucose	Platelet
Power	Other

The docking bays are located at appropriate spatial intervals across the vasculoid surface. Once a tanker docks, the manifest is scanned by a computer and the cargo is approved or rejected. Then, the empty tanker is simply reloaded or released empty into the traffic, depending on metabolic requirements at the time (remember, the tankers carry water, glucose, and gas). The whole process takes no more than 10 sec. The cellulocks, on the other hand, are embedded at various intervals across the vasculoid surface and are the sites at which the boxcars deliver biological cells. Because of the cargo they receive, the cellulocks are found in different densities throughout the vasculoid. For example, in the intestines, lungs, and throat, the cellulock densities are very high but regardless of the density, the function is the same: after securely docking with their loading face against the vasculoid surface, the boxcar doors open and the cellulock doors dilate to accommodate the opening, ensuring there is no leakage of material. At the end of the transfer cycle, the doors reseal and the boxcar undocks and continues on its way.

With billions and trillions of tankers and boxcars transporting cargo through the vasculoid, you might be thinking how everything is routed without getting lost. To avoid lost cargo, the outer surfaces of the tankers and boxcars (Table 10.1) bear a special interlock pattern akin to the Braille system (which can be reset depending on type of cargo). Similarly, the cilia surrounding each docking bay are fitted with special gripper pads that stick to the tanker and boxcar surfaces. It is a simple but effective recognition system that ensures the docking/undocking process is specific and timely.

It sounds like an amazing system, but surely there must be some drawbacks? Well, one of the boxcar's limitations is its size, since it cannot squeeze through the narrow capillaries found in the human retina, which may be as small as 4 microns wide. The vasculoid solves this minor design flaw by having the boxcars offload their cargo at cellulocks located outside tissues with narrow capillaries, leaving trans-tissue mobility to do the rest. To prevent the over-sized boxcars entering restricted passageway areas, they are simply rerouted, or prevented from entering by cilia equipped with gripper pads that are programmed to refuse boxcars.

Another important component of the amazing vasculoid is the vasculocyte. These are independent nanobots equipped with ambulatory appendages, manipulator arms, repair and assembly tools, onboard computers, communications, and independent power supplies. They patrol the vasculoid continuously, searching for maintenance and repair tasks such as plugging internal leaks, cleaning spills, leak scavenging, repairing/replacing malfunctioning ciliary leg mechanisms, clearing jammed docking bays, and even reconstruction of the vasculoid sapphire surface plate arrays (Figure 10.2).

Power and biocompatibility

By now, you may be wondering where the all energy comes from to power all these boxcars, nanobots, and tankers, but this is not a problem because the vasculoid is surrounded by a practically inexhaustible power supply: the chemical combination of oxygen and glucose, of which there is no shortage in the human body, since basal body power dissipation is about 100 W and 1600 W at peak exertion level.[2]

So far, it sounds like a neat system, but how can a human have such a device inside them without encountering compatibility issues? After all, it is an artificial system that is moving around the body. Well, obviously the vasculoid nanobots must be mechanically biocompatible with the vascular walls otherwise the blood vessels will become inflamed. One of the problems Freitas and Phoenix had to address was the shear stresses across the cell surfaces, especially in areas of bulk flow and in areas

---

[2]   At basal levels, the combined power requirement of the four subsystems requiring a regular energy supply (ciliary transport, docking bays, vasculocytes, and cellulocks) is about 27 W (193 W at peak levels).

Figure 10.2. Cell Rover. These nanobots patrol the circulatory system, searching for breaches. © 2000 E-spaces and Robert A. Freitas Jr, 3danimation.e-spaces.com and www.rfreitas.com and Phillippe Van Nedervelde.

where hydrostatic pressures were particularly high. One solution was to retrofit the patient with self-expanding vascular stents in areas such as the aorta, where flow and shear stress are especially high. The vascular stents are simply flexible Teflon or silicon coils or open-mesh tubes that are surgically inserted into an artery, expanded, and pressed into the vascular wall. Another concern was the effect the vasculoid's construction materials (diamond and sapphire) might have on a patient's immune systems but because of the biological inertness of the two materials, this is very unlikely. In fact, diamond nanoparticles have been shown to *enhance* immune system responses. If, in the unlikely event that the vasculoid did cause damage, the system would simply engage its self-repair mechanisms (this is a physiological process called angiogenesis, which involves the growth of new blood vessels from pre-existing vessels; the vasculoid would support angiogenesis by replacing damaged components from onboard inventories of spare parts).

System control and reliability

A system of 500 trillion units obviously requires a special control and communication system. In the vasculoid, this is achieved by special communications subsystems that ensure the plates are where they are supposed to be and cargo traffic is routed to where it is supposed to go. The vasculoid's complex system is also controlled by local computing centers located throughout tissues and a complex inter-plate network that enables communication with distant portions of the vasculoid network.

Now, installing a system as complex as the vasculoid might sound like a risky proposition. Surely, something so complicated cannot be reliable? Not so, say the designers, who seem to have thought of everything. To allay fears of system failure, the vasculoid's major subsystems incorporate 10-fold redundancy. That's right – 10 times system redundancy (many of the systems in the Space Shuttle, by comparison, have three levels of redundancy). Also, the elements of the vasculoid operate way below rupture strength, so structural failure is unlikely. If you recall, the tankers' rupture strength exceeds 40,000 atmospheres, yet they are loaded to only 1,000-atmospheres pressure and in the unlikely event of a rupture, the likelihood is the vasculoid's walls would contain the trauma.

Installation

By now, you may be wondering how a human could be fitted with such a device. Unlike donor organs, which are implanted, the vasculoid is installed in a complex process that begins with exsanguination and finishes with an intricate vascular plating operation. After being sedated, the patient's natural circulatory fluids are removed and replaced with installation fluids. This step is followed by mechanical vascular plating, defluidization, and finally activation of the vasculoid and re-warming of the patient. From start to finish, the installation process takes about 6 hr. What follows is the step-by-step process as it may occur in the near future.

A patient being installed is informed they are about to undergo a major medical procedure which involves replacing about 8% of the body mass with complex nanomachinery. Such an operation is not without risk, so prospective patients receive psychological counseling to deal with the personal implications. Preparation begins 24 hr before installation, when the patient receives an injection of 70 billion vascular repair nanorobots. These mobile, artery-walking nanobots clean out any fatty streaks, plaque deposits, lesions, infections, and vascular wall tumors. After completing their tasks, the repair devices are exfused and the results downloaded to a computer. This information is used by the surgeon to prepare a map of the patient's vascular tree to improve efficiency during plating and plate initialization. Once this step is complete, the patient is sedated, cannulated, and hooked up to a heart–lung machine, which supplies the equivalent of resting cardiac output through the femoral vessels allowing the entire human blood volume to be exchanged every few minutes. Heparin and streptokinase are then injected to prevent clotting, after which the surgeon administers various agents to aid the installation process. The patient is now ready for washout.

After the patient has been anesthetized, their entire blood volume is replaced with a suspension of respirocytes (spherical oxygen/CO_2 1,000-atmospheres-pressure vessels) and a mixture of electrolytes and other components normally found in blood substitutes. The respirocyte fleet provides oxygen and CO_2 transport equivalent to the entire human RBC mass for 3 hr after the cessation of respiration. Once the blood volume has been exchanged, the patient's core temperature is reduced from 37°C to just 7–17°C, after which the patient is ready for the intravenous deployment of vasculoid components. First, the respirocyte suspension is replaced by a new suspension containing 1% fully charged respirocytes and 10% cargo-bearing vasculocytes, creating a mixture whose viscosity and flow characteristics approximate to human blood. Each vasculocyte drifts in the flow until it encounters a vessel wall, which activates it, causing it to release its cargo. If the immediate area is already plated, the vasculocyte simply walks across the surface until it reaches a clear area to deposit its cargo. Once its cargo plate is in place, the vasculocyte releases back into the fluid, powers down, and is exfused from the body. After positioning and subsystem validation, each plate inflates fluid-tight metamorphic bumpers along its contact perimeter with its neighbors, which lock their bumpers firmly together with reversible fasteners embedded in the bumpers. After about an hour, the structure of the vasculoid is almost complete and all major components have been tested. The patient is now ready for defluidization.

During the defluidization stage, a monolayer of nanorobotic plates forms a chemically inert, flexible sapphire liner on the vascular tree's interior surface and vasculo-infusant fluid is purged from the body by introducing 6 l of oxygenated acetone to rinse the vascular tree. Once the system has been rinsed, the process of plate initialization begins. With 200 billion vasculocytes and 150 trillion plates to initialize, each active vasculocyte must contact and initialize 750 plates. This stage is followed by the installation of storage vesicles that contain reserves of mobile and cargo-carrying nanodevices and other auxiliary nanodevices. After a navel access port is installed, the tanker and boxcar population is introduced into the vasculoid and the patient is re-warmed, catheters removed, and the vascular breaches sealed. At this stage, the vasculoid is now operational and essential metabolic and immunological systems have returned to normal. While the vasculoid is intended to be a permanent feature, in the event of unforeseen circumstances, the system can be extracted if necessary.

A step closer to homo aquaticus?

Will the vasculoid bring us closer to realizing the science fiction writers' vision of homo aquaticus? By installing the vasculoid, humans will have the ability to breathe oxygen at very low partial pressures, since the gas tanker fleet can hold about 20 min of oxygen at the basal metabolic rate and up to 100 min if most of the tanker fleet is tasked with carrying respiratory gas. This means a human installed with a vasculoid would be capable of performing freedives for 1 hr or more, compared to the 5 or 6 min currently achievable by the best freedivers in the world today. It also means that scuba-divers breathing pressurized air would be immune to decompression sickness (DCS), since the vasculoid would simply rid the body of any excess nitrogen. Furthermore, vasculoid-installed scuba-divers would be able to dive to much greater depths without incurring any DCS penalty because the tanker fleet (capable of holding nitrogen at 1,000-atmospheres pressure, remember) would be capable of storing sufficient nitrogen to support excursions to 100 m or more. In fact, for divers intent on maximizing the advantages of the vasculoid, it would be possible to install an auxiliary nano-lung within the vasculoid wall, which could provide extra nitrogen storage capacity.[3]

Before you rush out to enquire about being installed with a vasculoid to improve your diving, the design described here is a provisional one at best and may not be realized for many years. The installation of such a device into a human for the purpose of extending the diving envelope may represent one of the most extreme interventions that will only be possible if significant advances in medical molecular nanotechnology are realized. However, current knowledge of nanomechanical systems suggests that such a device would not violate known physical, engineering, or medical principles and could be made safe for the user. If, in fact, the vasculoid becomes a reality, it may represent a significant outpost not only in biological evolution, but in Man's quest to extend the underwater frontier, and while it may not result in homo aquaticus, it will certainly bring us one step closer.

REFERENCES

[1] Freitas, R.A., Jr. Respirocytes: High Performance Artificial Nanotechnology Red Blood Cells. *NanoTechnology Magazine*, **2**(1) 8–13 (October 1996).

[2] Freitas, R.A., Jr. Exploratory Design in Medical Nanotechnology: A Mechanical Artificial Red Cell. *Artificial Cells, Blood Substitutes, and Immobilization Biotechnology*, **26**, 411–430 (1998).

[3] Freitas, R.A., Jr. Robots in the Bloodstream: The Promise of Nanomedicine. *Pathways, The Novartis Journal* **2**, 36–41 (October–December 2001).

---

[3]   Some divers may wonder how the nitrogen load would be extracted quickly enough to prevent DCS, since nitrogen extraction rates vary markedly by tissue type, but quick extraction would be achieved by installing the diver with a customized respirocyte-class nanobot.

Appendix

Since this book discusses aspects of diving that require a basic understanding of physics, I have included this appendix, which explains the basics of Henry's Law, the problems of dissolved nitrogen, "fast and slow" tissues, and a primer on decompression sickness (DCS). I hope it helps!

HENRY'S LAW

The amount of any given gas that will dissolve in a liquid at a given temperature is a function of the partial pressure of the gas that is in contact with the liquid and the solubility coefficient of the gas in the particular liquid. This is Henry's Law, but what does it mean? To understand this, go to the kitchen and pull out a Coke bottle from the fridge. Inside the bottle, there is the Coke and in a space at the top, just under the cap, there is a pressurized gas, which is carbon dioxide. Henry's Law states that because a liquid is in contact with a gas, that gas will dissolve into the liquid. The higher the pressure of the gas, the more of that gas will dissolve. That is why Coke is fizzy; pressurized carbon dioxide has dissolved into it! Now, to understand how Henry's Law affects scuba-divers, it is necessary to have a primer on basic physiology. We need air to breathe. Air comprises 79% nitrogen and 21% oxygen. Oxygen and nitrogen in the air travel through the lungs and pass into the bloodstream, where they are pumped in the blood around the body by the heart. Now, remember the Coke bottle? Replace the Coke with your blood and the carbon dioxide with air you breathe. The air divers breathe is pressurized and the deeper divers dive, the higher the pressure of this air/gas. The interface between the Coke and the carbon dioxide is in the diver's lungs. The deeper divers dive, the greater the pressure of the air he/she has to breathe and the more air is dissolved into the bloodstream. Now, we can divide the air into its constituent parts (nitrogen and oxygen) and we can refer to the *partial pressures* or ratios of each constituent. By doing this, we can understand the relative amounts of nitrogen and oxygen that dissolve into a diver's bloodstream because it is the dissolved nitrogen that gives divers the most problems. That is because the amount absorbed by the blood increases as divers go deeper for longer. Imagine a diver at 30-m depth. At this depth, the ambient pressure is 4 bar, which means four times the pressure of the air the diver was breathing at the surface! Now, some simple mathematics: 0.79×4 bar $= 3.16$ bar of nitrogen beginning to enter the diver's bloodstream compared to 0.79×1 bar $=$ only 0.79 bar entering at the surface. It is a big increase and all this nitrogen travels around your body and eventually is absorbed by the tissues in your body that your bloodstream services.

TISSUE SATURATION

Some body tissues absorb nitrogen quickly and some absorb it quite slowly. It is possible to determine the amount absorbed by different parts of the body by dividing the body up into compartments that absorb the gas at similar rates. The tissues that absorb nitrogen quickly are simply termed *fast tissues* and the ones that take longer are *slow tissues*. Fast tissues are ones such as the heart, lungs, and abdominal organs. These tissues become saturated within minutes, whereas the slower tissues, such as cartilage and joints, may take hours. This process of tissues becoming saturated is where the term *saturation diving* comes from. Divers can track how much nitrogen is in their system by using decompression tables (Figure A.1). Spending a long time at a deep depth is designated by a high letter in the alphabet denoting a high level of absorbed nitrogen. Now, you may be wondering how divers get rid of all that nitrogen. The answer is quite simple.

OFF-GASSING

Remember the Coke bottle? When you unscrew the cap, the Coke spills out of the bottle because by your unscrewing the cap, the pressurized carbon dioxide (the gas under the cap) has escaped. This is because Henry said that the amount of gas that will dissolve in a liquid depends on the pressure of the gas in contact with the liquid. By removing the cap from the Coke bottle, the pressure of the gas in contact with the liquid has been reduced. Since there is too much gas in the Coke, it simply comes out of solution. It is basically Henry's Law acting in reverse. Now, the faster the cap is unscrewed, the faster the pressure of the gas in contact with the liquid is changed and the faster the gas comes out of solution. So, if you unscrew cap *slowly* to allow the gas to come out of solution *slowly*, the gas will come out of solution slowly. But, if you see the gas coming out of the Coke too quickly, you would probably close the cap. Then you would wait a while before slowly opening it again. In fact, you might repeat this procedure two or three times to slowly release the gas. Sound familiar?

Now, let's see what's happening inside a diver when he/she ascends. We know he/she has high-pressure gas in contact with the bloodstream/tissues, which, during the dive, have been absorbing the gas. At this stage during the dive, the fast tissues would be totally saturated while the others would only be partially saturated. As the diver begins the ascent, he/she begins to reduce the pressure of the gas in contact

TABLE 1

Depth feet / metres		Doppler No-Decompression Limits (mnutes)	No-Decompression Limits and Repetitive Group Designation Table For No-Decompression Air Dives										
			A	B	C	D	E	F	G	H	I	J	K
10	3.0		60	120	210	300							
15	4.5		35	70	110	160	225	350					
20	6.0		25	50	75	100	135	180	240	325			
25	7.5	245	20	35	55	75	100	125	160	195	245		
30	9.0	205	15	30	45	60	75	95	120	145	170	205	
35	10.5	160	5	15	25	40	50	60	80	100	120	140	160
40	12.0	130	5	15	25	30	40	50	70	80	100	110	130
50	15.0	70		10	15	25	30	40	50	60	70		
60	18.0	50		10	15	20	25	30	40	50			
70	21.0	40			5	10	15	20	30	35	40		
80	24.0	30			5	10	15	20	25	30			
90	27.0	25			5	10	12	15	20	25			
100	30.0	20			5	7	10	15	20				
110	33.0	15				5	10	13	15				
120	36.0	10				5	10						
130	39.0	5				5							

Group Designation: A B C D E F G H I J K

Figure A.1. This a typical No-Decompression Limits (NDL) and Repetitive Group Designation Table used by divers that shows the maximum times a diver can stay at a set depth without having to perform a decompression stop. It also provides the group designator – a letter of the alphabet that designates the amount of nitrogen accumulated on the dive. The table also shows decompression stops at 10 feet, although this part of the table is for emergency use only.

with the bloodstream (like unscrewing the Coke bottle cap). The gas (we'll focus on nitrogen) in the diver's system has to be released as the pressure of the nitrogen in contact with the blood has been reduced, just like the carbon dioxide being released from the Coke. By utilizing safe ascent rates, the nitrogen is simply routed back into lungs and escapes as the diver exhales. But, if the diver ascends too fast, the nitrogen may come out of solution and form bubbles in the tissues/bloodstream. This often results in *decompression sickness* (DCS).

DECOMPRESSION SICKNESS

Bubbles are not supposed to be in the bloodstream. Gas is supposed to be in solution and as soon as bubbles form, the diver may be in trouble. Some bubbles, called *microbubbles*, are always present after every dive. In fact, it is possible for a diver to listen to them by hooking themselves up to a Doppler machine. However, the larger bubbles (the ones that cause DCS) are transported by the bloodstream and can cause all sorts of problems. For example, if a bubble is routed through a blood vessel for which it is too big (bubbles get bigger by expansion of the gas as the diver ascends), it will block off the flow of blood to that vessel and the tissues depending on that blood vessel for blood will be starved of oxygen and die. Bubbles also tend to cling together because of surface tension – a problem that can only be resolved by a visit to a decompression chamber, which reduces the size of the bubble(s) by forcing them back into solution. If the diver is lucky, he/she may survive without any long-term side effects. If he/she is unlucky and a bubble makes its way to the nervous system, the end result may be fatal.

SAFE ASCENTS

To avoid the insidious and potentially deadly effects of DCS, divers keep the bubbles in solution by ascending slowly. In fact, on some deep dives that heavily load the tissues, it is not possible to perform a continuous ascent. The diver must instead ascend a short distance and stop, wait for a while and then continue the ascent, repeating the procedure several times before reaching the surface. These stops are called *decompression stops* and the deeper a diver has dived, the more frequent the stops and the longer the intervals. Divers can gauge how much time they need to spend on decompression stops using the decompression tables mentioned previously.

Epilogue

"From birth, man carries the weight of gravity on his shoulders. He is bolted to earth. But man has only to sink beneath the surface and he is free."

Jacques Yves Cousteau

Hopefully, this book has provided readers with an insight into the challenges that lie ahead in venturing deeper into the ocean. At this juncture, the justification for moving forward into the ocean frontier is a combination of human, scientific, and economic issues and in anticipation of the challenges of establishing a more permanent human presence underwater, it is necessary we understand the steps that will take us there. Ultimately, the future of humans underwater will mirror the exploration of the space frontier and engage not just explorers, scientists, and engineers, but cooperation among private industry and tourists, too.

This book was written not only because the future of humans underwater merits telling, but also because there is a dearth of information describing how aquanauts will cope with the challenges of undersea living. Liquid breathing, artificial gills, and underwater outposts may be seductive concepts, and may even sound like science fiction to some, but for many of the technologies you have read about in this book, it is not a matter of decades, but a matter of years and, in some cases, months before these breakthroughs are realized.

In any environment in which air is absent, whether it is space or the deep ocean, humans will face challenges and risks but, as with any human enterprise with great risk, there is the possibility of great reward. There are no guarantees, but, in the future, people may remember the first few decades of the twenty-first century as a time when humans proved they were able to live for extended periods underwater.

Index